Luigi Troiano · Alfredo Vaccaro ·
Nishtha Kesswani · Irene Díaz Rodriguez ·
Imene Brigui
Editors

Progresses in Artificial Intelligence & Robotics: Algorithms & Applications

Proceedings of 3rd International Conference on Deep Learning, Artificial Intelligence and Robotics, (ICDLAIR) 2021

Springer

Editors
Luigi Troiano
Department of Management
and Innovation Systems
University of Salerno
Salerno, Italy

Nishtha Kesswani
Department of Computer Science
Central University of Rajasthan
Ajmer, Rajasthan, India

Imene Brigui
EMLYON Business School
Écully, France

Alfredo Vaccaro
Department of Engineering
University of Sannio
Benevento, Italy

Irene Díaz Rodriguez
Department of Computer Science
University of Oviedo
Gijón, Spain

ISSN 2367-3370 ISSN 2367-3389 (electronic)
Lecture Notes in Networks and Systems
ISBN 978-3-030-98530-1 ISBN 978-3-030-98531-8 (eBook)
https://doi.org/10.1007/978-3-030-98531-8

This Springer imprint is published by the registered company Springer Nature Switzerland AG
The registered company address is: Gewerbestrasse 11, 6330 Cham, Switzerland

Contents

An Opinion Mining of Text in COVID-19 Issues Along with Comparative Study in ML, BERT & RNN

Md. Mahadi Hasan Sany[1(✉)], Mumenunnesa Keya[1], Sharun Akter Khushbu[1], Akm Shahariar Azad Rabby[2], and Abu Kaisar Mohammad Masum[1]

[1] Department of Computer Science and Engineering, Daffodil International University, Dhaka, Bangladesh
{mahadi15-11173,mumenunnessa15-10100,sharun.cse, abu.cse}@diu.edu.bd

[2] The University of Alabama at Birmingham, Birmingham, USA
arabby@uab.edu

Abstract. The global world is crossing a pandemic situation where this is a catastrophic outbreak of Respiratory Syndrome recognized as COVID-19. This is a global threat all over the 212 countries that people every day meet with mighty situations. On the contrary, thousands of infected people live rich in mountains. Mental health is also affected by this worldwide coronavirus situation. Due to this situation online sources made a communicative place that common people shares their opinion in any agenda. Such as affected news related positive and negative, financial issues, country and family crisis, lack of import and export earning system etc. different kinds of circumstances are recent trendy news in anywhere. Thus, vast amounts of text are produced within moments therefore, in subcontinent areas the same as situation in other countries and peoples opinion of text and situation also same but the language is different. This article has proposed some specific inputs along with Bangla text comments from individual sources which can assure the goal of illustration that machine learning outcome capable of building an assistive system. Opinion mining assistive system can be impactful in all language preferences possible. To the best of our knowledge, the article predicted the Bangla input text on COVID-19 issues proposed ML algorithms and deep learning models analysis also check the future reachability with a comparative analysis. Comparative analysis states a report on text prediction accuracy is 91% along with ML algorithms and 79% along with Deep Learning Models.

Keywords: Covid-19 · Mental condition · Supervised learning · Depression · MNB · DTC · RFC · SVC · K-NN · LR · RNN · BERT

1 Introduction

Sentiment analysis these days can be considered as one of the foremost famous inquiries about themes within the field of characteristic dialect preparation. Sentiment analysis is

L. Troiano et al. (Eds.): ICDLAIR 2021, LNNS 441, pp. 1–16, 2022.
https://doi.org/10.1007/978-3-030-98531-8_1

built on the utilize of natural language processing, biometrics, substance examination, and computational historical underpinnings to proficiently bring out, degree, recognize and dismember passionate conditions and personal data [1]. The employments of estimation examination are secured by a few curiously logical and commercial ranges, such as supposition mining, event detection, and recommender frameworks [2]. These days the social media stages such as Facebook, Twitter, WeChat, Instagram and YouTube, are an extraordinary field of data known as social information [3] besides that many people express their opinion on other platforms like different websites and survey administration software like Google-forms, etc. Other than that the rise in emphasis on AI strategies for literary analytics and NLP taken after the colossal increment in open reliance on social media like Twitter, Facebook, Instagram, blogging, and LinkedIn for information, rather than on the customary news organizations [4–6]. Coronavirus (COVID-19) which started to seem after 2019 in Wuhan, China has been one of the foremost examined and one of the foremost overthrowing infections all over the world. By the result of World Health Organization (WHO) until we are composing this think (2021/4/27) more than 14, 68, 41,882 people have contracted the malady and more than 31, 04,743 people have passed on in 223 countries over the world (WHO | World Health Organization). People's lives are getting to be exceptionally dubious day by day as there's no treatable antibody. All through the history of mortality, social orders have been gone up against with crises of diverse sorts and of exceptionally distinctive natures, such as crises caused by the organization, send out of imperativeness resources, or emergencies caused by the impact of sicknesses and plagues, charitable clashes, and cash related crises, and among others [7]. We are in an exceedingly globalized world, where improvement flexibility, free travel between nations, and the headway and utilization of Data and Communication Developments (ICTs) are completely set up [8, 9]. But within the show circumstance, the current crisis caused by COVID-19 is making a social mechanical situation that's exceptional for society [10]. In expansion, we must highlight the numerous countries that are locked down to urge a diminished rate of infection and death and many people are stuck in their houses. And these situations are responsible for many pyromaniacs. In this study, we've attempted to analyze the feelings or sentiments of Bangladeshi people in this emergency circumstance using several supervised machine learning (ML) algorithms and deep learning (DL) methods utilized to prepare our model. We scarcely found any work utilizing this kind of Bangla data-set on sentiment analysis. There are parcels of emotion analysis related work with Facebook, Reedit, Twitter, and many other social media stages data except for raw data collection from general people by Google Form. We've collected parts of Bangla feedback from 443 general people related to covid-19. In our input shape contain two sorts of questions. Within the to begin with address, there are six sorts of suppositions where there are three positive and three negatives, and within the moment address depressed or not depressed. People gave criticism based on those six suppositions and against those conclusions, they moreover gave the input that whether they are depressed or not. Authorities ought to know people internal conditions so that authority can persuade them. We have moreover compared all algorithms and models for a way better understanding. The abbreviations we have used in our work is given bellow (Tables 1 and 2)-

Table 1. ML acronyms.

Acronym	Definition
MNB	Multinomial Naive Bayes
RFC	Random Forest Classifier
DTC	Decision Tree Classification
SVC	Support Vector Classifier
KNN	K-Nearest Neighbours
LR	Logistic Regression
RNN	Recurrent Neural Network
BERT	Bidirectional Encoder Representations from Transformers

Table 2. Acronyms for statistical measurement.

Acronym	Definition
CM	Confusion Matrix
MAE	Mean Absolute Error
MSE	Mean Square Error
RMSE	Root Mean Square Error
TP	True Positive
TN	True Negative
FP	False Positive
FN	False Negative
LL	Log Loss

2 Literature Review

Sentiment analysis can be a computational bridge to people's internal assumptions or sentiments. Previously, numerous studies have been carried out on sentiment analysis, to name a few, prediction on political approaches through political articles, speeches [11], opinions of news entities on newspapers and blogs [12], stock exchange [13, 14], cricket related comments from several social platforms using Bangla Text [15], etc. By utilizing data mining calculations, people's sentiment can be judged through social media. Recently, several groups of research scholars have devoted their effort to study and analyze information that is collected from numerous social media for different purposes using sentiment analysis. For this analysis many of them applied machine learning algorithms or deep learning models or both. Since the beginning of the widespread use of Covid-19, the work displayed by Aljameel et al. [16] pointed to conducting sentiment analysis on Saudi Arabia state's people of awareness on the Twitter-based data set. Nair et al. [17] they use BERT, VADER, and LR to find the sentiment of tweets whether it is positive or negative or neutral. In [18] Sabbir et al. they analyze on 1120 Bangla data using CNN and RNN to get output that is angry, depressed, or sad. Where CNN is able to get the most accurate and the model predicts that most people commented or expressed analytically. In [19] Shaika and Wasifa did sentiment analysis on Bangla Microblogs post. Rashedul et al. [32] approach an automated system where the system can categorise six individual emotion classes from Bangla text using supervised learning technique. In this pandemic situation, the Bangladesh government locked down properly to control the situation like other countries of the world. In this situation, people's mental situations are not good enough [20]. Most of the researchers collect their data from many social media and other sites, but we collect our data from general people by Google Forms where most of them are students from different universities at the age between 18–30 years age. Recently, Anjir et al. [33] used traditional ML algorithms and deep learning models such as RNN, RNNs, GRUs, BERT to analyze the sentiment of people about vaccination of Covid-19. They survey among the general people and data was collected by Google form. Galina et al. [34] also collect raw data by Google form to analyze the effect of Covid-19 pandemic on universities students learning.

We've utilized two classes to characterize the public's opinion which was collected by Google form. We've utilized a few machine learning algorithms and deep learning models for their better accuracy in our data set which is within the Bangla language.

3 Methodology

The outbreak of the Corona epidemic has had a huge impact around the world, especially in the population and among the elderly, adults, and middle-aged people. It has caused enough fear and anxiety in the world. So we did a survey of how people are spending their time in lockdown during this Covid-19 situation and what their mental state has been like while sitting at home. The methodology part is used to identify, select, process, and analyze information on any one subject in the study. In this part, we discuss in detail the research methodology such as - whether the qualitative or quantitative method of research has been used or why (Fig. 1).

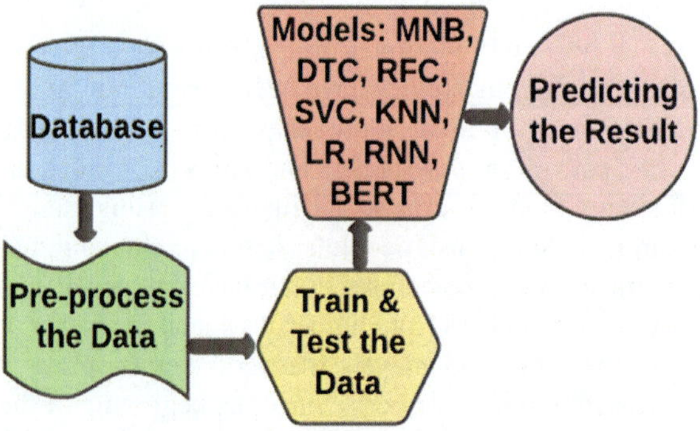

Fig. 1. Proposed model of our work [21].

3.1 Data Collection

Data is the most important and vital part of Artificial Intelligence. The more we will give data the more machines can predict and analyse impeccably [22]. A machine's efficiency depends on the quality and the sum of information. In this proposed approach we've collected data from the general people using social media like Facebook and messenger chat groups by Google Form shown in Fig. 2. We were able to take 1085 responses based on 2 questions where the first question contains 6 opinions where 3 are positive and 3 are negative and based on these opinions, everyone responds in the second question whether they are depressed or not shown in Table 3. Finally, we've got two columns 'TypeOfOpinion', and 'Status' where 'TypeOfOpinion' refers to the opinions and 'Status' refers to the status of depression level. Among all of those responses of the moment address, we've found 550 under depressed, 535 not depressed.

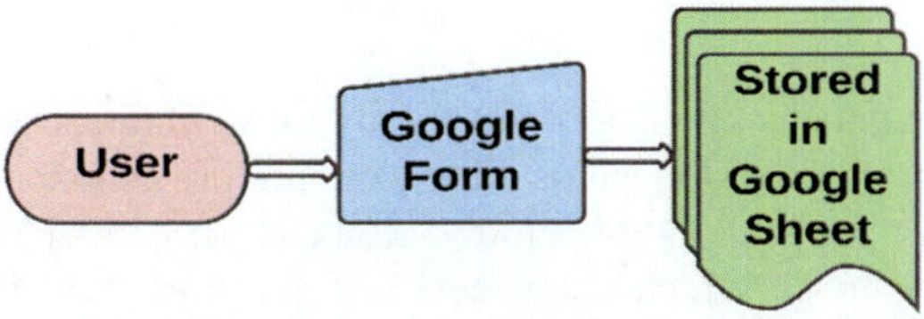

Fig. 2. Google form.

Table 3. Responses.

কভিড-১৯ এর সুবিধা এবং অসুবিধা (একাদিক অপশন সিলেক্ট করতে পারবেন) [Advantages and disadvantages of Covid-19 (you can select multiple option)] as TypeOfOpinion	**আপনি কি মানসিকভাবে অবসাদগ্রস্ত ? [Are you mentally depressed?] as Status**
স্বাস্থ সচেতন হয়েছি, এই সময়ে সময়ে এক্সট্রা এক্টিভেটিস করার যথেষ্ট সময় পেয়েছি (Having become health conscious, I have had enough time to do extra activities from time to time)	Yes
স্বাস্থ সচেতন হয়েছি (I became health-conscious)	No

3.2 Pre-processing

Data preprocessing is required tasks for cleaning the data and making it reasonable for a machine learning show which moreover increases the accuracy and productivity of a machine learning model because real-world information generally contains commotions, lost values, and possibly in an unusable arrangement which cannot be specifically utilized for machine learning models [23]. In preprocessing there are two parts one is preprocessing for ML algorithms and preprocessing for DL methods. Before applying ML algorithms, we did some preprocessing steps, at first, the "Status"-column attributes "yes" and "No" were considered to be "1" and "0" respectively. Then in the data set there are a lot of punctuation used which is needless like as “[(), $ %^&*+={ }\[\]:\"|\'\~‘ <]”.

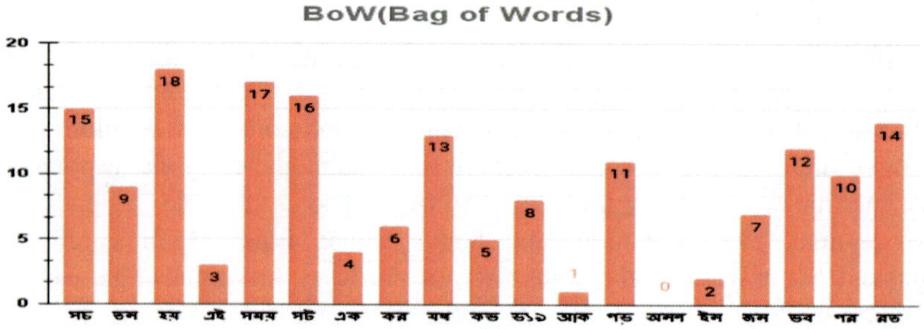

Fig. 3. BoW

We remove this punctuation systematically by using libraries. After removing the punctuations, we remove the stop words. Stop word implies commonly used words like “[অতএব, অথচ, অথবা, অনুযায়ী, অনেক, অনেকে, অনেকেই, অন্তত, অন্য]”and so on. We’ve utilized the Bangla data set. Which had parcels of stop words. Those have been removed technically [24]. Then we utilize Bag of Words (BoW) in our data. BoW is utilized to change over a collection of content archives to a vector of term/token counts. It also empowers the pre-processing of content information earlier to produce the vector representation. This usefulness makes it a profoundly adaptable include representation module for content. After using BoW on my data set the segmented part for “স্বাস্থ সচেতন হয়েছি, এই সময়ে সময়ে এক্সট্রা এ এক্টিভেটিস করার যথেষ্ট সময় পেয়েছি, কভিড- ১৯ এ আক্রান্ত হয়েছি, পড়ালেখা বিঘ্নিত হয়েছে”this sentence is like below Fig. 3. As like as ML here, we did some pre-processing step for feeding our RNN model, we implement tokenization. Tokenization fundamentally alludes to part up a bigger body of content into littler lines, words or indeed making words for a non-English dialect. The different tokenization capacities are in-built into the ‘nltk’ module itself and can be utilized in programs. For our data set, we use word tokenization to split the sentences into smaller pieces like: 'পড়', 'সময়', 'হয়'[25]. After that we labelencode the data set. LabelEncoder can encode a single variable. In this part, after tokenizing the whole data we encode the train and test part of the “Status” column by using LabelEncoder so that our model can do the further process for predicting accurately. Before feeding the model for train data BERT model expects input data in a specific format. In the BERT model “Tokenizer” does the preprocessing by itself. All the vectors ought to be of the same measure. Consequently, it should cushion the content to realize a common length. This length will be decided by the length of the longest sentence, which we’ll need to calculate. Too, since we should concatenate the TypeOfOpinion and the Status column, the length ought to be chosen by the concatenated content column. Then tokenized the dataset that means splitting a sentence into a smaller as like: 'পড়', 'সময়', 'হয়,which is described above in the ‘RNN’ part. Including extraordinary tokens to check the starting ([CLS]) and separation/end of sentences ([SEP]). Then it creates token IDs. Pad the sentences to a common length. Create attention masks for the above PAD tokens.

3.3 Training Parameter

In Table 4, the table is demonstrated for all information that is individually liable for the result which we actually got in result,

Table 4. Parameter settings.

Algorithms	RNN			BERT
Train 90%	Train 75%	Beta-1 0.9	Epoch 15	Train 80%
Test 10%	Test 25%	Beta-2 0.999	Verbose 2	Test 20%
Random size 0	Learning rate 0.0005	Batch size 52	Validation split 0.15	Epoch 3

Here, In the algorithm column, we took train size 90% and the test size was 10% cause our data set was small that's why we took the train size 90 so that the machine can learn more from the data set. And we took random state 0 cause if the value of the random state zero then the machine will take it as much as it needs. In the RNN column, it's noticeable that the train size is 75% and the test is 25%. For the RNN model, we split the data set 75:25 for getting better accuracy. Besides that, the learning rate, beta1, and beta2 are liable to get the accuracy that we demonstrated in Table 6. In Table 7 we can see that in epoch 13 we got 94% validation accuracy when verbose was 2 and validation split was 15%. The batch size was 52 which means per epoch 332 (train size)/52(batch size) = 6.38 which means 6 steps will appear. Lastly in the BERT column train size is 80% and the test size is 20%. And the epoch size is 3, where 74 steps in per epoch.

3.4 RNN Model

Recurrent neural networks (RNN) is a sort of artificial neural network that includes extra weights to the network to make cycles within the network chart in an exertion to preserve an inside state. The promise of including state to neural networks is that they will be able to expressly learn and misuse setting in arrangement expectation issues, such as issues with an order or temporal component. In convolutional neural networks (CNN), all the inputs and outputs are autonomous of each other, but in cases like when it is required to predict the following word of a sentence, the last words are required and consequently, there's a have to be kept in mind the previous words. In this way, RNN came into existence, which solved this issue with the assistance of a Hidden Layer. The most vital feature of RNN is hidden, which recollects a few data almost an arrangement and RNN are a sort of Neural Network (NN) where the output from the last step is fed as input to the current step. RNN has a "memory" that recollects all data around what has been calculated. It employs the same parameters for each input because it performs the same errand on all the inputs or covered up layers to deliver the yield. This decreases the complexity of parameters, not at all like other neural networks.

3.5 BERT Model

BERT stands for Bidirectional Encoder Representations from Transformers. It is designed to pre-train significant bidirectional representations from the unlabeled substance by together conditioning on both left and right settings [31]. As a result, the pre-trained BERT show can be fine-tuned with a fair one extra output layer to make state-of-the-art models for a wide extent of NLP assignments. The BERT design builds on the best of Transformer. It has two variant right now, which are, BERT base & BERT large. Then it starts preprocessing with three combinations of embedding's which are Position Embedding, Segment Embedding and Token Embedding.

3.6 Statistical Analysis

Confusion Matrix: A confusion matrix may be a technique for summarizing the execution of a classification algorithm. Calculating a confusion matrix can give you a better idea of what a classification model is getting right and what sort of mistake it is making. The number of correct and erroneous predictions are summarized with tally values and broken down by each lesson. This can be the key to the confusion network. From confusion matrix we will able to get some values of precision, recall, and F1 score and equations are given bellow-

$$\text{The equation of precision} = \frac{True\ Positive}{True\ Positive\ +\ False\ Positive} \quad [26]$$

$$\text{The equation of recall} = \frac{True\ Positive}{True\ Positive\ +\ False\ Negative} \quad [26]$$

$$\text{The equation of F1 score} = 2 * \frac{recision\ *\ recall}{precision\ +\ recall} \quad [27].$$

4 Result and Discussion

This section will discuss all outcomes of both algorithms and the model part. Overall, we have worked on the humanoid condition of people during the Covid-19 period for which we have conducted a survey to find out about their human exhaustion. Through that survey, we found out if they are suffering from humanoid depression. And we are also aware of how they have benefited during the Coronation period.

ML Algorithms - Considering the bar chart in Fig 4, it is seen that great results have been obtained by applying machine learning models (e.g. MNB, RFC, DTC, SVC, KN, and LR) for the survey. And it can be seen that among all the algorithms RFC gives 91% result which is more than other algorithms.

As we can see in Table 5, some steps of the Confusion Matrix are illustrated, between the states "0 and 1" of the dataset which indicates whether the person is mentally exhausted or not. Also shown is the status of the six algorithms, the values of precision, recall, F1 çalculation, and support for "0 and 1", which are seen differently for

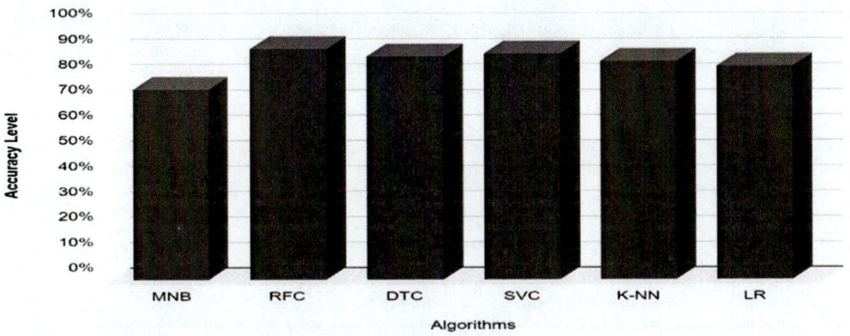

Fig. 4. ML algorithm accuracy bar chart.

different algorithms. But in the case of "0", it has the same support in each algorithm and has given another value for "1" but it is the same for all.

As we can see in Table 6, the macro and weighted averages of the Confusion matrix are calculated for each algorithm for precision, recall, F1 score, and support. These values differ between the six algorithms, and in some cases, it returns the same value. As we can see, in the case of MNB, it has given the same values of Precision, F1 score, and support, but in the case of Recall, it has given different values, again in RFC, DTC, and SVC, it has given the same value to all. On the other hand, it has given different scores of precisions and F1 on KNN where it is different in terms of recall and support. All showed the same value except the F1 score on the LR. But for our proposed model SVC did better than K-NN wherein [28] K-NN did better than SVM.

Here, Table 7 demonstrates the False Positive (FP), False Negative (FN), Negative Prediction Value (NPV), False Discovery Rate (FDR), Mean Absolute Error, Mean Squared Error, Root Mean Squared Error, Root Mean Squared Error, and Log Loss. We have shown these scores in this table just for the six algorithms and in the case of the model, only Bert shows all the scores of the Confusion matrix as shown in Table 10, 11.

Sensitivity and Specificity of all algorithms which we use for our dataset. Sensitivity is used in Fig. 5 because it can predict the outcome of variable decisions over a certain range. It creates specific sets for variables and can determine the analyst by which it influences the result through a change. It also uses sensitivity analysis to determine the effects of variables. The specification has been used only to increase the flexibility of the models. In the x-axis, we have represented the algorithms and placed the levels towards the y-axis.

Figures 4, 5 and Table 5, 6, 7, show all the features when training size 0.9 and test size 0.1. The results that come after changing the standard of training and examination are summarized in Table 5 through table. When the values of training and testing were 0.9 and 0.1, the accuracy of the model was much higher which is very low in the case of Table 5. We wanted to change the quality of training and testing over and over again to see how accurate the models are. As a result, when we compare Fig. 4. And Table 8, we can see how much the quality of training and testing varies in terms of accuracy.

DL Models - This part describes all of the experimental results for our RNN and BERT models. After applying both models with the data set accuracy and loss result of train

Table 5. Result of precision, recall, F1-score, support.

Algorithms	Status	Precision	Recall	F1-score
MNB	0	0.87	0.59	0.70
	1	0.70	0.91	0.79
RFC	0	0.91	0.91	0.91
	1	0.91	0.91	0.91
DTC	0	0.87	0.91	0.88
	1	0.91	0.87	0.89
SVC	0	0.90	086	0.89
	1	0.88	0.91	0.89
K-NN	0	0.86	0.86	0.86
	1	0.87	0.87	0.87
LR	0	0.94	0.73	0.82
	1	0.79	0.96	0.86

Table 6. Macro and weighted average for precision, recall, F1-score, and support.

Algorithms	Macro and weighted average	Precision	Recall	F1-score
MNB	Macro avg.	0.78	0.75	0.75
	Weighted avg.	0.78	0.76	0.75
RFC	Macro avg.	0.91	0.91	0.91
	Weighted avg.	0.91	0.91	0.91
DTC	Macro avg.	0.89	0.89	0.89
	Weighted avg.	0.89	0.89	0.89
SVC	Macro avg.	0.89	0.89	0.89
	Weighted avg.	0.89	0.89	0.89
K-NN	Macro avg.	0.87	0.87	0.87
	Weighted avg.	0.86	0.87	0.86
LR	Macro avg.	0.86	0.84	0.84
	Weighted avg.	0.86	0.84	0.85

and test are shown in Table 9. Table 9 is the comparative study with the six algorithms of Fig. 4. Table 9 assumes the RNN & BERT model train, test accuracy, loss, and all accuracy. We can see that the ML models give the best accuracy rather than the DL models. Between the two models BERT performs the best, it's both train test accuracy is higher than RNN and also the loss is lower than the other model.

Table 7. Confusion matrix for all algorithm.

Algorithm	FP	FN	NPV	FDR	MAE	MSE	RMSE
MNB	0.3	0.13	0.91	0.41	0.24	0.25	0.49
RFC	0.08	0.09	0.91	0.09	0.08	0.08	0.29
DTC	0.09	0.13	0.86	0.09	0.11	0.11	0.33
SVC	0.12	0.09	0.91	0.13	0.11	0.11	0.39
K-NN	0.13	0.14	0.86	0.14	0.13	0.13	0.36
LR	0.21	0.05	0.95	0.27	0.15	0.15	0.39

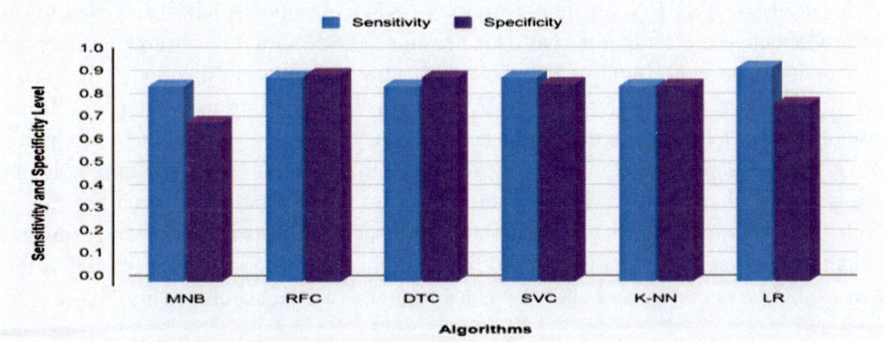

Fig. 5. Sensitivity and specificity bar chart.

Table 8. Comparative accuracy for ML algorithms.

Train/Test Size	MNB	RFC	DTC	SVC	K-NN	LR
Train = 0.50 Test = 0.50	72.97	68.01	68.47	67.12	69.82	68.92
Train = 0.60 Test = 0.40	67.42	69.01	73.03	66.85	74.16	67.98
Train = 0.70 Test = 0.30	74.44	74.44	73.68	67.70	75.19	67.70
Train = 0.75 Test = 0.25	75.68	76.58	73.87	65.77	77.48	64.87
Train = 0.80 Test = 0.20	74.15	79.78	78.65	67.41	69.66	67.41

Table 9. Accuracy and loss of RNN and BERT model.

Models	Train		Test		All accuracy
RNN	0.72	0.54	0.71	0.55	72%
BERT	0.73	0.53	0.53	0.51	79%

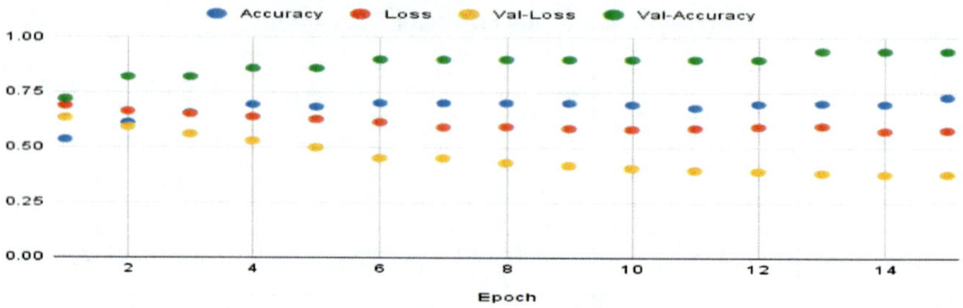

Fig. 6. RNN model History.

Model History of RNN is demonstrated in Fig. 6, 15 epochs have been used to run the model because after that it gives static results. That's why it has been kept in epoch 15. And it can be seen from Fig. 6 that over time the "accuracy" of the model, the "loss", the "validation accuracy" and the "validation loss", are getting more and less. In the case of RNN, we have got a maximum accuracy of 72% with a validity of 94% which is for very good modeling. From Fig. 6, it's noticeable that in the "Loss" column reducing rate is slower than the "Val-Loss" column. On the other hand, the increasing rate of "Val-Accuracy" is higher than "Accuracy". Figure 7 represents the ratio of training and validation accuracy and loss graph. From the graph, we can see that the accuracy is less than validation accuracy and also loss is less than the model accuracy (Fig. 8).

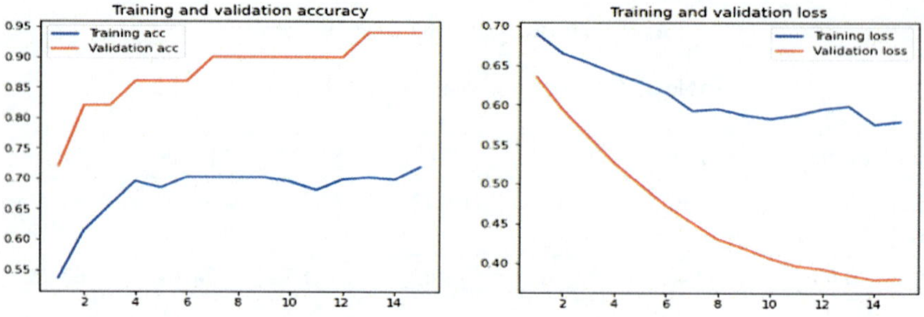

Fig. 7. Graphical representation of RNN

The scenario is like column "Loss". In columns "Accuracy" and "Val-Accuracy", there the 2nd epoch value is lower than the 1st epoch and the 3rd epoch is higher than both the 1st and 2nd columns, and the whole scenario is different from the Table 8 which is for the RNN model.

Here, Table 10 representing the Precision, Recall, F1-Score, and Support of "Status". The resultant value describes that for "1" the precision is higher than "0" and recall is the opposite of precision. F1-Score and Support result difference is almost the same. Comparing Table 5 BERT gives almost a lower result for Classification report because BERT always works on big data but our data source is too short to run the model but still, it performs better in the small dataset.

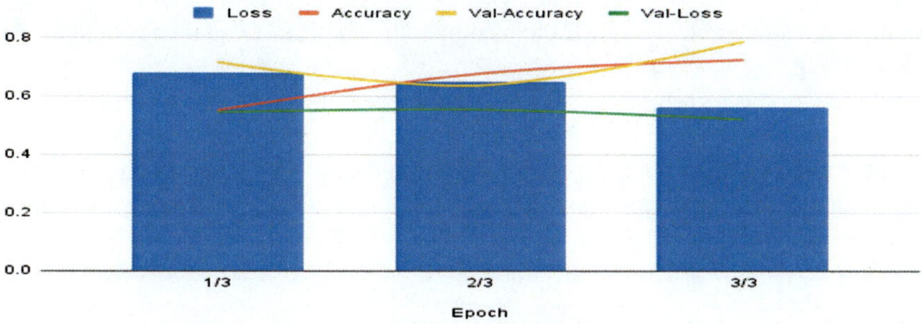

Fig. 8. BERT model History.

Table 10. Precision, recall, F1-score, and support.

Model	Status	Precision	Recall	F1-score
BERT	0	0.72	0.86	0.78
	1	0.87	0.73	0.79

Table 11. Precision, recall, F1-score.

Model	Macro and weighted average	Precision	Recall	F1-score
BERT	Macro average	0.79	0.80	0.79
	Weighted average	0.80	0.79	0.79

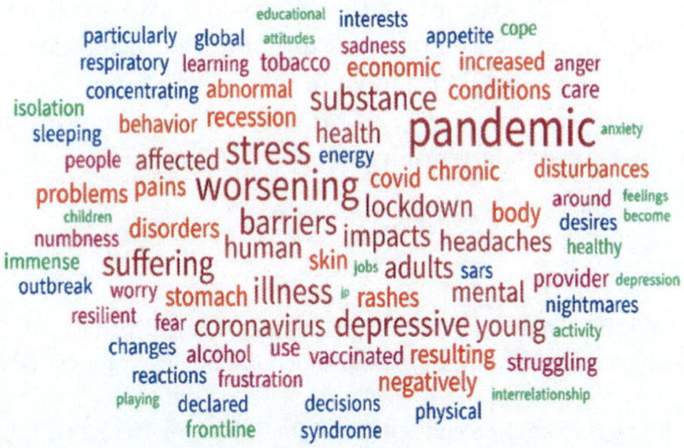

Fig. 9. Word cloud of mental health during Covid-19 pandemic.

Support of macro and Weighted average from the BERT model. As in Table 11 the weighted average is higher than the Macro average for precision and Recall is the opposite of precision and the F1-Score and Support result difference is the same Macro &

Weighted average. Table 11 and Table 6 comparatively almost equal and for short data, it's a big thing.

A word cloud is a grouping of words that are represented in various sizes. The larger and bolder the phrase, the more often it appears in a text and the more significant it is. In Fig. 9, the word cloud is a mental health condition during the Covid-19 situation where the most significant words of this topic are mentioned and the words collect from our dataset showed in the word cloud.

5 Conclusion and Future Work

From this experiment, the proposed model can predict the sentiment of a Bangla sentence whether it's depressed or not. This model is proposed on only 443 data which was collected by Google form. Here we applied several supervised machine learning classification algorithms and also deep learning models which were RNN and BERT. Be-fore applying these algorithms and models some preprocessing techniques and some heterogeneous features to feed our algorithms and models. Then we got 71% accuracy on deep learning-based model RNN and 79% on BERT besides that on algorithm portion RFC achieved 91%, SVC 89%, DTC 88%, K-NN 86%, LR 84%, and MNB 75% which is the lowest among all algorithms and RFC is the highest among models and algorithms. This experiment will conclude that by the victimization of this different kind of machine learning classification algorithm we will get better results, and con-jointly using the deep learning models achieved very high accuracy even in an exceedingly touch of training data. We have some circumscription that the whole experiment was did on 443 data. And the data set was based on 6 opinions and there were 2 categories against the opinions whether it is depressed or not depressed. In the future, we'll experiment with totally diverse preprocessing strategies, heterogeneous features, super-vised and unsupervised calculations for creating an extra exact framework and also work with social media related data. For stirring giant data; we will use projected heterogeneous options and deep learning options like Word2Vec and Word Embedding to apply deep learning algorithms like Long short-term memory (LSTM) [30] and Convolutional neural networks (CNN) [29] to induce outstanding results.

References

1. Kumar, M., Anju, B.: Analyzing Twitter sentiments through big data. In: 2016 3rd International Conference on Computing for Sustainable Global Development (INDIACom). IEEE (2016)
2. Sailunaz, K., Reda, A.: Emotion and sentiment analysis from Twitter text. J. Comput. Sci. **36**, 101003 (2019)
3. Priyanka, T., Tripathi, R.C.: A review towards the sentiment analysis techniques for the analysis of twitter data. In: Proceedings of 2nd International Conference on Advanced Computing and Software Engineering (ICACSE) (2019)
4. Neri, F., Aliprandi, C., Capeci, F., Cuadros, M., By, T.: Sentiment analysis on social media. In: 2012 IEEE/ACM International Conference on Advances in Social Networks Analysis and Mining (2012). https://doi.org/10.1109/asonam.2012.164

5. Alexandra, B.: Sentiment analysis in social media texts. In: Proceedings of the 4th Workshop on Computational Approaches to Subjectivity, Sentiment and Social Media Analysis (2013)
6. Yue, L., Chen, W., Li, X., Zuo, W., Yin, M.: A survey of sentiment analysis in social media. Knowl. Inf. Syst. **60**(2), 617–663 (2018). https://doi.org/10.1007/s10115-018-1236-4
7. Bentley, J.H.: The Oxford Handbook of World History. Oxford University Press, Oxford, UK (2012). ISBN 9780191744051
8. Guidry, J.P.D., Jin, Y., Orr, C.A., Messner, M., Meganck, S.: Ebola on Instagram and Twitter: how health organizations address the health crisis in their social media engagement. Public Relat. Rev. **43**(3), 477–486 (2017). https://doi.org/10.1016/j.pubrev.2017.04.009
9. James, J.: Globalization, Information Technology, and Development, Palgrave Macmillan UK, London, UK (1999). ISBN 978-1-349-40631-9
10. Gao, J., et al.: Mental health problems and social media exposure during COVID-19 outbreak. Plos one **15**(4), e0231924 (2020)
11. Young, L., Soroka, S.: Affective news: the automated coding of sentiment in political texts. Polit. Commun. **29**(2), 205–231 (2012). https://doi.org/10.1080/10584609.2012.671234
12. Godbole, N., Srinivasaiah, M., Skiena, S.: LargeScale Sentiment Analysis for News and Blogs. ICWSM 2007 Boulder, Colorado, USA (2007)
13. Liang-Chih, Y., Jheng-Long, W., Chang, P.-C., Chu, H.-S.: Using a contextual entropy model to expand emotion words and their intensity for the sentiment classification of stock market news. Knowl.-Based Syst. **41**, 89–97 (2013). https://doi.org/10.1016/j.knosys.2013.01.001
14. Hagenau, M., Liebmann, M., Neumann, D.: Automated news reading: stock price prediction based on financial news using context-capturing features. Dec. Supp. Syst. **55**(3), 685–697 (2013). https://doi.org/10.1016/j.dss.2013.02.006
15. Wahid, M.F., Hasan, M.J., Alom, M.S.: Cricket sentiment analysis from bangla text using recurrent neural network with long short term memory model. In: International Conference on Bangla Speech and Language Processing (ICBSLP) (2019)
16. Aljameel, S.S., et al.: A sentiment analysis approach to predict an individual's awareness of the precautionary procedures to prevent COVID-19 outbreaks in Saudi Arabia. Int. J. Environ. Res. Public Health **18**(1), 218 (2021)
17. Nair, A.J., Veena, G., Vinayak, A.: Comparative study of Twitter Sentiment on COVID-19 Tweets. In: 2021 5th International Conference on Computing Methodologies and Communication (ICCMC), pp. 1773–1778. IEEE (2021)
18. Pran, S.A., Bhuiyan, R., Hossain, S.A., Abujar, S.: Analysis of Bangladeshi People's Emotion During Covid-19 in Social Media Using Deep Learning. IEEE – 49239 (2020)
19. Shaika, C., Chowdhury, W.: Performing sentiment analysis in Bangla microblog posts. In: 2014 International Conference on Informatics, Electronics & Vision (ICIEV). IEEE (2014)
20. Kontoangelos, K., Economou, M., Papageorgiou, C.: Mental Health Effects of COVID-19 Pandemia: A Review of Clinical and Psychological Traits. Published online 2020 Jun 15. https://doi.org/10.30773/pi.2020.0161, PMCID: PMC7324731, PMID: 32570296
21. Masum, A.K.M., Khushbu, S.A., Keya, M., Abujar, S., Hossain, S.A.: COVID-19 in Bangladesh: a deeper outlook into the forecast with prediction of upcoming per day cases using time series. Procedia Comput. Sci. **178**, 291–300 (2020). https://doi.org/10.1016/j.procs.2020.11.031
22. Roh, Y., Heo, G., Whang, S.E.: A survey on data collection for machine learning: a big data-AI integration perspective. IEEE Trans. Knowl. Data Eng. **33**, 1328–1347 (2019)
23. Kalra, V., Aggarwal, R.: Importance of text data preprocessing and implementation in RapidMiner. In: ICITKM (2017)
24. Kanakaraj, M., Guddeti, R.M.R.: NLP based sentiment analysis on Twitter data using ensemble classifiers. In: 2015 3Rd International Conference on Signal Processing, Communication and Networking (ICSCN). IEEE (2015)

25. Pak, A., Paroubek, P.: Twitter as a corpus for sentiment analysis and opinion mining. In: LREc, vol. 10, no. 2010 (2010)
26. Derczynski, L.: Complementarity, F-score, and NLP Evaluation. In: Proceedings of the Tenth International Conference on Language Resources and Evaluation (LREC'16), pp. 261–266 (2016)
27. Govindaraju, V., Zhang, C., Ré, C.: Understanding tables in context using standard NLP toolkits. In: Proceedings of the 51st Annual Meeting of the Association for Computational Linguistics (Volume 2: Short Papers) (2013)
28. Huq, M.R., Ali, A., Rahman, A.: Sentiment analysis on Twitter data using KNN and SVM. Int. J. Adv. Comput. Sci. Appl. **8**(6) (2017)
29. Huang, Z., Xu, W., Yu, K.: Bidirectional LSTM-CRF models for sequence tagging. arXiv:1508.01991 (2015)
30. Alam, M.H., Rahoman, M.M., Azad, M.A.K.: Sentiment analysis for Bangla sentences using convolutional neural network. In: 2017 20th International Conference of Computer and Information Technology (ICCIT). IEEE (2017)
31. Clark, K., Khandelwal, U., Levy, O., Manning, C.D.: What does bert look at? An analysis of bert's attention. arXiv:1906.04341 (2019)
32. Tuhin, R.A., Paul, B.K., Nawrine, F., Akter, M., Das, A.K.: An automated system of sentiment analysis from Bangla text using supervised learning techniques. In: 2019 IEEE 4th International Conference on Computer and Communication Systems (ICCCS), pp. 360–364. IEEE (2019)
33. Chowdhury, A.A., Das, A., Saha, S.K., Rahman, M., Hasan, K.T.: Sentiment Analysis of COVID-19 Vaccination from Survey Responses in Bangladesh (2021)
34. Ilieva, G., Yankova, T., Klisarova-Belcheva, S., Ivanova, S.: Effects of COVID-19 pandemic on university students' learning. Information **12**(4), 163 (2021)

AI-ML Based Smart Online Examination Framework

Swapnil Sapre(✉), Kunal Shinde, Keval Shetta, and Vishal Badgujar

Department of Information Technology, A.P. Shah Institute of Technology, Thane, India
{swapnilsapre,kunalshinde,kevalshetta,vsbadgujar}@apsit.edu.in

Abstract. Online examinations are the way of conducting examinations on the user's mobile devices or laptops rather than actual paper. In situations when physical exams cannot be conducted online exams have been the preferred choice. In physical examinations, the number of students doing malpractices reduces by a great amount as there are examiners physically present to monitor every student. The idea of Smart AI-ML-based Online Framework is to reduce the malpractices done by the students in the online mode as far as the current platforms in use are concerned. For this, there are some methods used that are based on machine learning algorithms. The main task during the exam time is the continuous live detection of every candidate and for this facial detection using a webcam is necessary. Before this, a proper pose setup and facial illumination need to be ensured. With the live detection happening the examiner can track the student from their end as well as get the alerts based on the student's illegal conduct. Overall, the system can perform the proctoring tasks in a fully automated fashion and thereby requiring very minimal effort from the examiner in monitoring the candidates. We have concluded that the Smart Online Examination platform is a much viable solution to the existing platforms for conducting the exams and doing the proctoring.

Keywords: E-learning · Online proctoring · Face detection · Artificial intelligence

1 Introduction

The traditional examination system process is long and time-consuming whereas the online examination system provides a quick and accurate solution within the specified time frame. Online exams require fewer resources and are more reliable and accurate as compared to offline tests. Students have quick access to online exams thereby reducing the necessity for faculty and staff for managing a large number of teaching materials and dealing with seating arrangements. The system provides a fast and accurate solution within the specified time frame and the scoring process is simpler and saves teachers time. The assessment is quick, reliable, and accurate. The online exam provides flexibility and security within it, as each student can receive random questions from an equivalent sample test. The online exam allows one to add videos, images, audio, PDF files, and more to the tests. Online exams allow for several features to stop malpractices, like locking the browser, disabling the print, arrow, and ESC keys [2].

L. Troiano et al. (Eds.): ICDLAIR 2021, LNNS 441, pp. 17–25, 2022.
https://doi.org/10.1007/978-3-030-98531-8_2

Proctored exams are a type of online exam to protect the integrity and prevent malpractices of students. When such an online examination system is combined with automated proctoring it adds more value to the entire system as the teacher can conduct exams in a much efficient manner by getting all the activity of candidates at one place [4, 7]

1.1 Facial Recognition and ML

Facial Recognition is a process that maps an individual's face and saves it as a face print. The software program makes use of algorithms to look at a captured face and compare it to the saved one to verify one's identity. Machine learning algorithms best recognize the numbers so it is pretty challenging. This numerical illustration of a face is named a function vector. As a result, using machine learning the function vector can detect the following aspects in a person's face:

1. Height of face (cm)
2. Width of the face (cm)
3. Average color of face (R, G, B)
4. Width of lips (cm)
5. Height of nose (cm)

By combining machine learning and making it applicable for online examinations has done betterment in this field. They yield a system that is fully automated in conducting examinations, monitoring each candidate without the examiner needing to do it individually, and grading the students along with getting automated attendance while also maintaining the records for future reference [6].

The distribution of this paper from hereafter is as follows: Sect. 2 contains the literature survey done and Sect. 3 presents the objectives of the project. Section 4 describes the existing system architecture, methods, and the solution to that is shown in the proposed system architecture in Sect. 5. Section 5.1 contains the results of the proposed system along with the strength and weaknesses it contains. Section 6 contains the conclusion.

2 Literature Survey

In [1] authors have discussed the facial feature detection algorithm which is needed to detect the face of the candidate from several image sets. It is based on the average variance calculation of the three components in the face image like eyes, nose, and mouth and the extraction of these features.

In [10] authors have discussed CNN (Convolutional Neural Networks) a deep learning algorithm that helps in face verification and detection. The system requires training samples by taking a large number of sample data of students' faces obtained from online lec tures and examinations.

In [5] authors have discussed a tab locking mechanism that ensures candidates do not navigate away from the exam screen because their exam window is locked while the exam is in progress. By having randomized system for question papers the students do not get the same questions and the questions are also shuffled.

YOLO is a machine learning algorithm that is used to perform object detection. It can be used to detect objects in still images, provided sample video, or from a real-time webcam which is used in this system as discussed in [9]

Deepface is a deep learning python-based framework that can detect faces from images. It is a neural network-based and has already been trained with sample images of other people.

To capture the candidate's eye position, eye tracking and pupil detection are necessary. Authors in [8] present face landmark detection which is a computer vision-based method to detect and track the user's eye movement by forming key points on it.

3 Objectives

1. To set up proctoring mechanism to stop malpractices in an exam by creating a system that validates each user well before the exam using facial recognition.
2. To allow the faculty to alert warnings or end the exam if students are found performing malpractices.
3. To provide a platform for the faculty to set the questions and also have support for multiple languages and detailed reporting.
4. To allow faculty to give additional times to students with disabilities
5. To send attendance and activity report details to faculty
6. To deploy our framework on the cloud to achieve load balancing and scalability.

4 Existing System Architecture

There are many tools and software available on the market for taking tests online. However, most of them are hosted inside the system and are not intended for free use. Monitoring is not enabled on other widely used open-source evaluation platforms. This allows students to engage in professional illicit activity, such as talking to others or minimizing the screen to find answers while taking the test. The existing system includes the following aspects:

1. The system includes a simple username and password combination to log into the exam portal (Fig. 1).

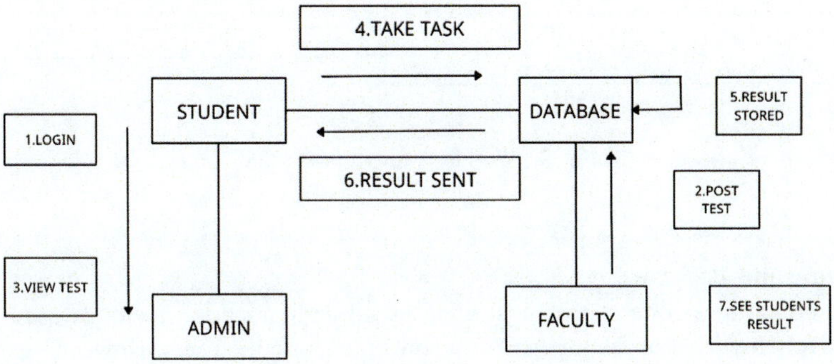

Fig. 1. Authentication of users

2. For taking the attendance, the faculty has to supervise the student by calling out their name.
3. There is no diversity in questions as there are a series of similar questions available in the questionnaire for all the students taking the exam.

This project aims to solve the limitations of the existing systems and further develop a new system.

5 Proposed System Architecture

The proposed system is explained by dividing it into two blocks of sequence diagrams, one for the student end and another one for the examiner end.

Student Block. Before the candidate begins to attempt the examination they must register and login into the exam portal with their credentials.

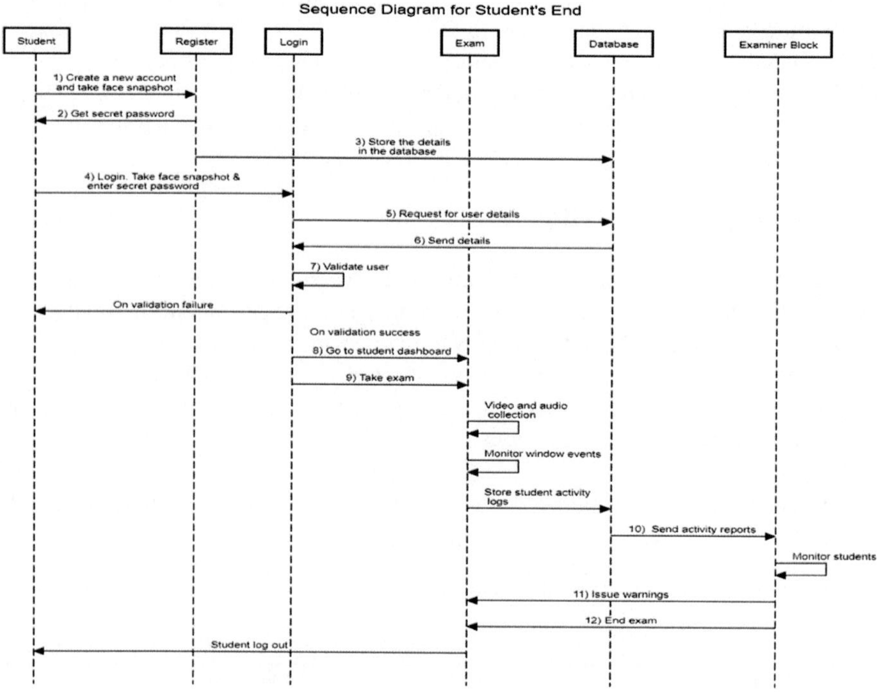

Fig. 2. Workflow for student's end

Username and Password

This combination is one of the most widespread authentication methods used due to its relatively low cost of deployment and convenience to the users. However, users can impersonate another person and there would be no means to detect if an online student is the one he or she claims to be. For this purpose, two-factor authentication mechanism is used. During the registration process, the student's mobile number or email plays a very important role. Every time the student logins for an exam an OTP (one-time-password) will be sent to the registered mobile number or email. As a result, one student cannot take the exam on the other's behalf as one student can register only with one mobile number/email at a time.

Figure 2 illustrates the process followed by the system from the students' perspective. The student taking up the examination needs to set up a profile in advance. The profile will contain all the personal information along with the pictures of them taken by varying distances and light variations. For this, the system requires the use of a webcam of the device to take pictures of the user's face. On the day of the examination, the students have to enter their username as they are recognized during the registration process along with some other documents as proof [3]. Upon successful login, they are directed to the facial recognition page. New images of the user are taken and compared against the ones stored in the database to establish if they match within suitable bounds. Once the student is validated by this method, they are allowed to proceed with their examinations. This completes the two-factor authentication process as now the students are verified both by secret password and face recognition.

Examiner Block. Figure 3 below illustrates the process followed by the system from the examiner's perspective.

The examiner will be using this system for exam generation and monitoring the students while they are giving the exam. Another feature the proposed system provides is the ability to issue warnings to the students doing malpractices and eventually ending the exam for those who are repeatedly found doing the same. The exam generation process and the activity report details methods are demonstrated in the next subsections.

Question Paper Generation Process. The most crucial task that the examiner needs to carry out is the creation of the question paper. The system has been developed to increase diversity and avoid all the students getting the same questions and in the same order [5].

At the time of creation the examiner can specify the type of question from different options provided as per their choice.

The proposed system makes all examiners mandatory to display the questions in the form of images rather than plain text. This ensures that students do not copy-paste the text as it is in another tab to get the answers. The system also allows the examiner to create a CSV/doc file-based question paper and then simply import it in.

Alert Generation. Alert is an important feature implemented in this system. The webcam and microphone are used to capture the live images and audio from the student's end. The examiner is provided with a dashboard that enables them to see live data of all the students. Monitoring each individual is not possible in a real-time scenario and the alert based system generates such alerts. The alerts are displayed to the examiner containing the activity carried out.

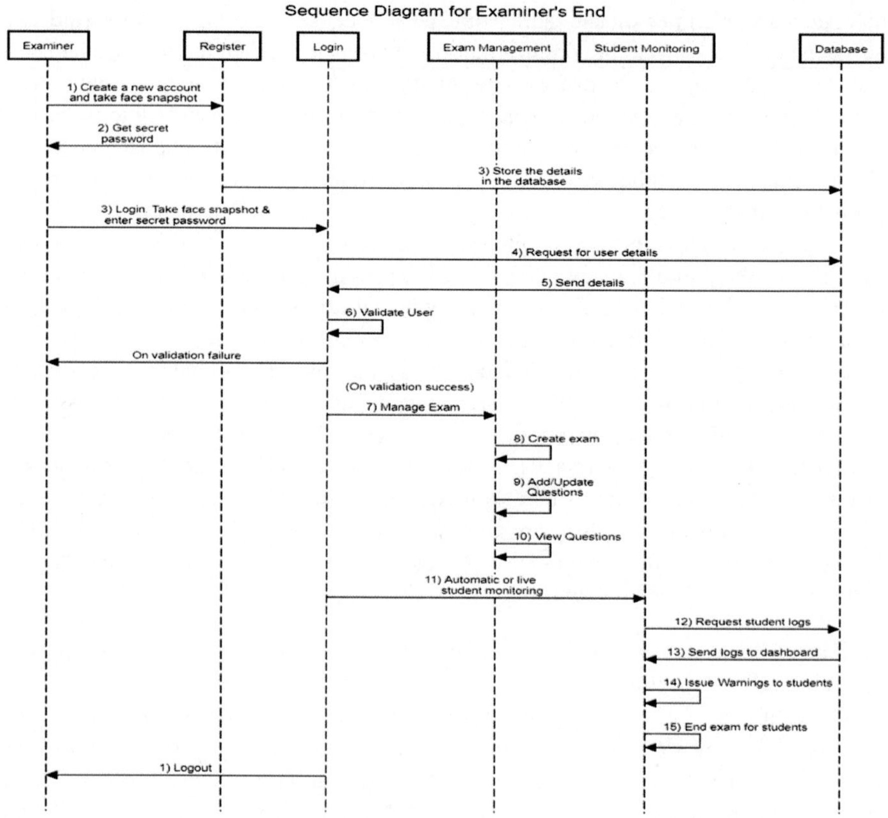

Fig. 3. Examiner's end process

Based on the above report, the examiner can send warnings to the student. After issuing multiple warnings if the student is still found doing malpractices then the examiner can decide to end the exam for that particular student.

The next section discusses some of the methods used to detect the face, video, and audio of the candidates

Object Tracking and Face Recognition. Dlib is a C++ based library consisting of machine learning algorithms and trained data sets that are capable of performing face recognition. In the centroid tracking algorithm, to detect any object a unique ID has to be assigned to it and then track in subsequent frames. This requires every object to be associated with a unique ID which will take up memory and is not a feasible solution. For this reason, we are adopting the correlation tracking method which is as follows:

i) detect the object once or in specific time intervals
ii) Apply the algorithm to track the object in subsequent frames.

The machine learning algorithm makes use of predefined data sets to track the object. For the dlib tracker to track the same object we need to define the boundary around our target specified in four coordinates (top, right, bottom, left) which are passed to a predefined function 'rectangle' as shown below.

rect = dlib.rectangle(startX, startY, endX, endY)

The sample input is passed to the dlib object tracker which tracks it in further frames. The updating process of the tracker is performed at the back end. The process will be looped throughout the examination process to track the students.

The candidate will be allowed to give the examination only when they are verified by using face recognition and validation. The OpenCV library of python is capable of processing the images and videos to identify the faces and objects that appear in the samples. The face recognition module method used generates a 128-d real-valued number feature vector per face. This is like assigning unique codes to each face.

Figure 4 shows how the sample picture has been associated with the 128-d values. Once this has been accomplished, the encoded value and corresponding name from the database are assigned to it. In the actual process of examination, the images of candidates are detected, and then a box is applied around the face of the person (as marked by a white rectangle in the image above) based on the four coordinates and then convert this into the 128-d number value.

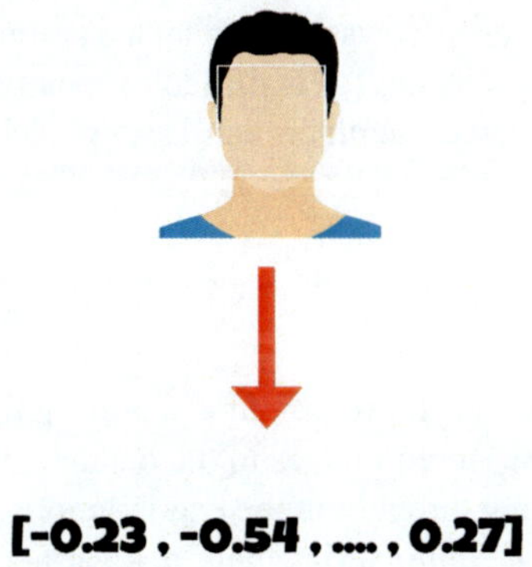

Fig. 4. Generating 128-d real number per face

We attempt to match each face in the input image to our encoding in the database. The result of this process will be as follows:

The distance is within an acceptable value indicating the faces match or else the faces do not match. The object tracking algorithm also performs the process of object detection in the candidate's live environment simultaneously. The object detection happens in the video frames in combination with the YOLO real-time object detection algorithm [9].

Tab Locking. The last thing the system prevents the students from doing is accessing other windows to fetch the answers. Authors in [5] use a method by having the examination window locked until the exam is over and generating warnings to the student in case the student navigates to other windows by some other methods.

5.1 Results

The strengths of the proposed system include the two-factor authentication which is the combination of secret password and the face validation at the time of login for both the examiners and the students, log generation that gives a detailed report of all the activities done by the students during their examination on the examiner's dashboard, and efficient exam generation process for the exam creation and its management (Fig. 5).

(a) Face snapshot during registration (b) Login by a different user

Fig. 5. Face registration and validation process

The weaknesses of the proposed system are that the system can detect all the window and user movement-related events but it is difficult to analyze the front view from the candidate's perspective. Also, the examiners can't set up a subjective and practical-based examination.

6 Conclusion

In our system, face detection is achieved by first capturing the face snapshot of the user and then verifying it with the stored images in the database. The system is also capable of performing student tracking throughout the examination. As the input facial image is captured in real-time during examinations there is a variance in image quality and the illumination of the face. On this aspect, the YOLO algorithm has provided very accurate results by detecting the objects in the candidate's live environment. The designed system is also able to automate the attendance process by generating the data of the students who have appeared for the exam. Our examination framework currently handles only objective-based questions. Hence the future scope of this work is to make the entire system a part of the e-learning framework and also provide support for practical as well as subjective-based examinations.

References

1. Abd-Elhafiez, W.M., Heshmat, M., Elaw, S.: Efficient Method for Face Recognition and Its Role in Supporting E-Learning Systems (2015). https://doi.org/10.1109/ECONF.2015.21
2. Aimin, W., Jipeng, W.: Design and Implementation of Web-Based Intelligent Examination System (2009). https://doi.org/10.1109/WCSE.2009.77
3. Asep, H.S.G., Bandung, Y.: A Design of Continuous User Verification for Online Exam Proctoring on M-Learning (2019). https://doi.org/10.1109/ICEEI47359.2019.8988786

4. Atoum, Y., Chen, L., Liu, A.X., Hsu, S.D.H., Liu, X.: Automated online exam proctoring. IEEE Trans. Multimedia **19**(7), 1609–1624 (2017). https://doi.org/10.1109/TMM.2017.2656064
5. Chua, S.S., et al.: Online Examination System with Cheating Prevention Using Question Bank Randomization and Tab Locking (2019)
6. Damale, R.C., Pathak, B.V.: Face Recognition Based Attendance System Using Machine Learning Algorithms (2018). https://doi.org/10.1109/ICCONS.2018.8662938
7. Jegatha Deborah, L., et al.: Secure Online Examination System for e-learning (2019). https://doi.org/10.1109/CCECE43985.2019.9052408
8. Kainz, O., et al.: Enhancing Attention through the Eye Tracking (2020). https://doi.org/10.1109/ICETA51985.2020.9379229
9. Mahendru, M., Dubey, S.K.: Real Time Object Detection with Audio Feedback using Yolo vs. $Yolo_v3$ (2021). https://doi.org/10.1109/Confluence51648.2021.9377064
10. Sukmandhani, A.A., Sutedja, I.: Face Recognition Method for Online Exams (2019). https://doi.org/10.1109/ICIMTech.2019.8843831

Spaced Repetition Based Adaptive E-Learning Framework

Amit Kharwal, Neelay Umrotkar, Varun Godambe(✉), Uttam Kolekar, and Vishal Badgujar

Department of Information Technology, A.P. Shah Institute of Technology, Thane, Maharashtra, India
{amitkharwal,neelayumrotkar,varungodambe,udkolekar, vsbadgujar}@apsit.edu.in

Abstract. With global education moving towards online education, we see a lot of emerging techniques to help facilitate paradigm shift. Memorization being a fundamental part of the process of learning, we focus on making memorizing an efficient task for students. We plan to make use of 2 emerging methodologies that help us memorize and retain information better in the long term. These are: Spaced Repetition: this technique helps in retaining information efficiently. Adaptive Learning: This will help us make the process of learning personalized, so that the user is always challenged just enough to enable growth. To help users stay consistent, we plan to add game-like elements to the system to help facilitate regular use which will lead to better results.

Keywords: Spaced Repetition · Adaptive Learning · Memorization · Re-Collection · Flashcards · Machine Learning

1 Introduction

The Indian education system inherently places a lot of importance on memorization. Tests are designed to be subjective that require long descriptive answers which in turn need the user to either understand the topic to the core or memorize sets of information which they can string together during the test. This pattern is visible in throughout a student's learning career. Often one of the things that happen most often is that students rote-memorize information haphazardly and fail to reproduce it in the tests, and even if they do reproduce it properly, they do not recall it later on after the exams. This is a sub-optimal and incorrect way to go about testing and memorization/learning in general. We intend to develop a system that helps students memorize and retain information in a systematic and efficient way. In order to do this, we decided to use techniques like Spaced Repetition, Adaptive Learning and in conjecture with these, we plan to add game like elements to keep the users' interest so as to not miss out on practice. Below, we've explained how each of these techniques mentioned above will play a part in our project.

L. Troiano et al. (Eds.): ICDLAIR 2021, LNNS 441, pp. 26–34, 2022.
https://doi.org/10.1007/978-3-030-98531-8_3

Spaced repetition is a technique for efficient memorization which uses repeated reviews of content following a schedule determined by the spaced repetition algorithm to improve long term retention. We review information in the form of cards of information, at gradually increasing intervals. For example, if you answer a card correctly, the algorithm will repeat that card less frequently, whereas if you get the answer wrong, it'll be asked more frequently. The reason this technique works better is that the revision is done at or around the exact time when the user is expected to forget said information. Spaced Repetition is significantly more effective than rote learning [11].

Adaptive learning instead of a static and rigid learning system which forces everyone to follow the same learning pattern, you're given a personalized experience. If the learner takes a test, the algorithm can adapt the training according to the answers given. Doing badly on certain questions can trigger the system to push relevant content to the learner, or the overall performance could determine which learning path they take. Adaptive learning refers to measuring and responding to each activity while a student progresses in the material assigned by their instructor. When combined with data from student interactions, it is able to present the best content (right type, right amount, right level) for that moment in time to keep the student engaged. Flexible adaptive learning can help preserve the content regardless if you're teaching on campus, online or a mixture of both. Rooted in several learning science principles like spacing, chunking, and interleaving, create mini-cycles of questions that create smaller, easier to absorb chunks of content. If students continue to struggle with a concept, they are more likely to encounter repeating questions. This method can help instructors understand their students better by understanding the data provided by the quiz results [6]

Learning Curve: Two of the most common methods we see in the market are:

1) Pimsleur Method: Here you are asked to retain information learned after a predefined interval, without taking into account your performance in the previous review. This method is extremely restrictive and one-dimensional.
2) Leitner Method: Here, a few different intervals in terms of boxes, are predefined. Let's say we have the intervals of 1, 2, 4, 8, and 16 d. Every new piece of information starts in the 1 -Day Interval Box. If the user remembers the information after 1 d, the information gets put in the 2 - Day Interval Box. If the user remembers the information after 2 d, it gets promoted to the 4-Day Interval Box, so on and so forth. If at any moment, the user fails to recall the information, then the piece of information gets shifted to the previous interval box. This method was a massive upgrade over the Pimsleur Method. But it can still be made even better by using another method.

2 Objectives

1. To design an adaptive algorithm that takes user performance into account in order to adapt the next question's levels.
2. To design a space repetition algorithm that takes user's specific quiz performance into account.
3. To integrate Adaptive and Spaced repetition algorithms for the proposed system.
4. To add game design elements to make the layout appealing.
5. To create a cross-platform application interface.

3 Literature Review

In literature [4], a centralized system is proposed which will act as a portal for both students and instructors where the instructors can curate content according to their student's needs and the students can access this content. The performance of students on using this content will then be available to the instructors to then modify their offered content. The constant cycle of content modification according to the needs of the student and analysis of the student's performance greatly interest us. Keeping content relevant will be an important part of the memorization process and so will the analysis.

In literature [5], The literature ponders over how our traditional learning systems could implement a cloud-computing based e-learning system. It discusses the advantages of cloud-computing based architecture such as high availability which is definitely a value that we hold important in our own implementation.

In literature [6], the author focuses a significant chunk of itself on explaining the deference between personalized learning and adaptive learning, where, adaptive learning focuses on individual differences, individual performance and adaptive adjustment whereas personalized learning focuses on all of these but with the addition of personal needs. This here, makes a big difference. As our system focuses more on memorization and learning rather than teaching, one of the things we can adopt from this research is a feedback system which helps specialists and teachers cater to the student's needs better.

In literature [7], our takeaway is to increase the amount of content available for the user thereby reducing the chances of repetition in the same session while also removing the chances of the user cramming all the content of the application in one go, thereby maintaining the integrity of the spaced repetition-based system.

In literature [8], the author focuses on the effects of adaptive learning in the domain of language learning and teaching. In adaptive learning, learner's interaction with previous content determines subsequent content. Hence a sort of personalization is seen. Digital flashcards are used to incorporate automated spaced repetitions of the targets learning item. Since the system is individualized, it can adapt relatively easily in response to small amounts of data. Stated to be a 'hot concept' that is 'poised to reshape education' in Webley magazine. As we intend to make the application based on digital flashcards, we can use the idea of creating algorithms to determine the order, frequency and difficulty of the next learning item based upon their previous exposure to it.

In literature [9], the author provides and details an adaptive learning platform which was implemented from scratch and used for undergraduate courses related to digital systems and computers. Students differ in their learning capabilities and it will be difficult to provide a tailored education, especially in large classes. Different technology solutions have been proposed to deal with this challenge. Rule based approaches use pre-determined branching architectures based on his response to questions. If a question is answered correctly, the system could issue a more challenging question which is pre-determined. The system can issue an easier question or hints or other assistance if the question is answered incorrectly. Algorithmic mechanisms use mathematical models for inferring student's ability level perform real-time content adaptation. This paper helped us reinforce that adaptive learning does indeed help students memorize and learn better.

In literature [11], the author's model captures what the student has learned once a module is completed, and estimates how well he can recall this knowledge at any given time. Spaced repetition is a key component of the student model. Over time, the strength of a skill will decay in the student's long-term memory, and this model helps the student manage his practice schedule. The spacing effect is people remembering things more effectively if they use spaced repetition as opposed to cramming. We are given a brief exposure to primitive spaced repetition methods such as the Pimsleur Method and the Leitner Method. The main research that happens in the paper is related to the Ebbinghaus Forgetting Curve[10] and how the researchers tweaked the original curve to make a model that more performs better when calculating the knowledge half -life. We will be using the Ebbinghaus Curve model to has our Half-life calculations on.

In literature [12], the author provides us with a primitive idea of how spaced repetition helps students perform better and how gamification helps to have students engaged during the process of learning and how it also helps them return to the process regularly which in turn also helps us understand just how effective spaced repetition is. We do plan to implement game-like features in our project to showcase similar results of engaging and retaining students to the learning process. We would also, as is regularly done with any spaced repetition program, expose students to maintenance and acquisition cards (Figs. 1 and 2).

4 Existing System Architecture

Anki Spaced Repetition Tool: [2] SM-2 stands for Super Memo 2 [1], which is a spaced repetition algorithm. It was developed my P.A Wozniak. It has multiple versions right from SM-0 to SM-18. Despite these versions, SM-2 remains the most popular. In its calculations, it takes into account factors such as the "repetition number," "easiness factor" and "interpretation interval." Here, repetition number refers to how many times a certain element has been repeated. Easiness factor refers to how quickly the interval between each repetition increases. Inter-repetition interval refers to the amount of time between repetitions.

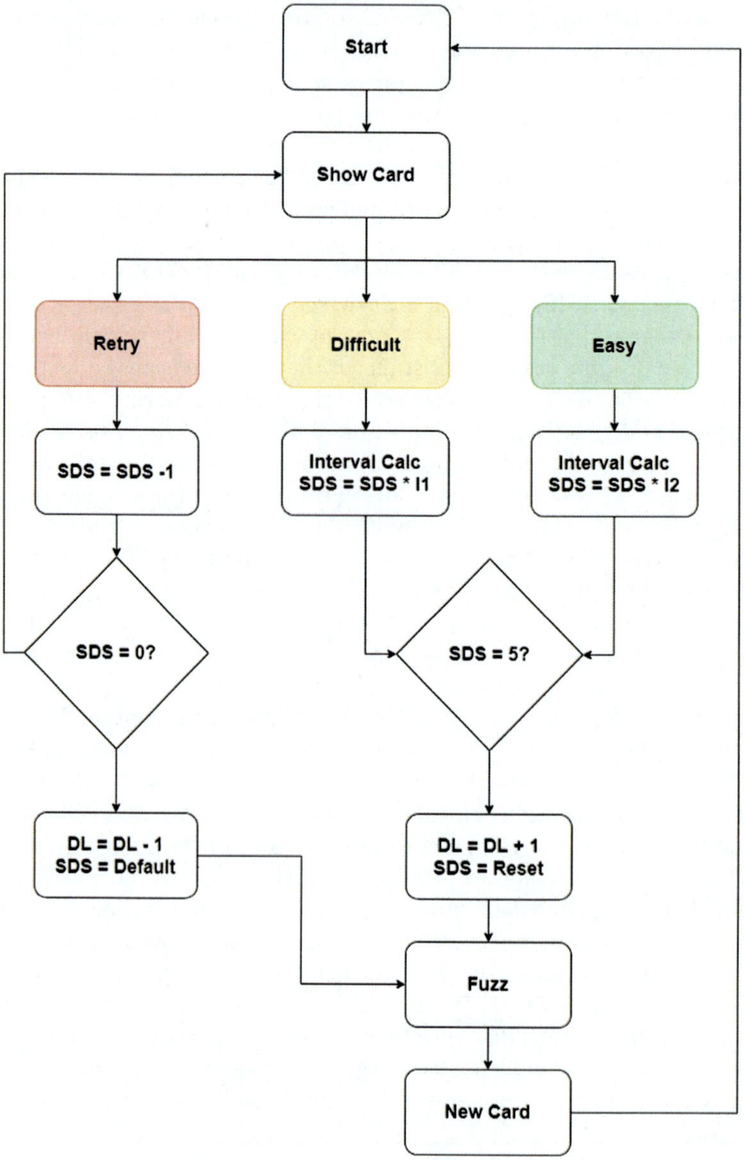

Fig. 1. Existing system architecture.

Legend	Meaning
E	Ease coefficient
I	Interval
EB	Ease Bonus Coefficient
Fuzz	Small Randomized Addition to Interval to Prevent Repetition of Card Sequence

Fig. 2. Legend for existing system architecture.

In Anki, they use a tweaked version of the SM-2 spaced repetition algorithm. Once the flashcard is showed on the screen, the user is given 4 options. If the user selects the "Repeat" option, the card is repeated again after the deck is done, and the Ease is decreased by 20% points. If the user selects the "Hard" option, the Interval is multiplied by 1.2 and the ease is increased by 15 percent points. If the user selects "Good" option, the interval is multiplied by ease, the Interval remains the same. If the user selects the "Easy" option, the Interval is multiplied by Ease * Easy bonus and Ease is increased by 15% points. Here. It is made that if the person is having difficulty with a certain question, it is repeated more often. The minimum time interval is never less than a day.

Although Anki covers the basics of spaced repetition very well, it lack interest building elements. To cover this, we looked at another language learning tool available on devices as well as on the web called "Clozemaster" [3]. Clozemaster is a tool that helps people learn and memorize vocabulary, but their approach to the same is very different. Where Anki uses the same only cards with a very toned-down visual environment, Clozemaster uses a lot of game-like elements to keep the user's interest. Although they don't used a proper spaced repetition system, the visual language that they use is something that can be learnt from. Elements that we found extremely necessary and helpful:

- Colorful Design
- Progress Bars
- Daily Goals
- Personal Stats
- Leader boards

5 Proposed Architecture

In our enhanced Spaced Repetition based Adaptive Learning Framework, we start of with the user being showed a flashcard. The user has been assigned default values of SDS and Interval. And the questions have the value of DL. Once shown the flashcard, the user has 3 options (Fig. 3):

Retry: If the user selects this option, first, his SDS is reduced by 1. Next it is checked if the SDS is equal to Zero. If it is, it means that the user hasn't been able to answer 5 questions of the same difficulty. In short, the questions are too difficult for him. In this case, the DL is reduced by one to ease the user and the SDS returns to 1. Now, new cards of the reduced difficulty are produced.

Difficult: Here, if the user selects this option, the interval is calculated using our half-life calculations based on Ebbinghaus' Forgetting Curve equations and the SDS is increased by 0.5.

Easy: Here, if the user selects this option, the Interval is again calculated using the half-life formula and the SDS is increased by 1.

In both Difficult and Easy options, if the SDS is equal to 5, the DL is increased by 1 thus producing new questions that are more challenging in nature to keep the user challenged.In our implementation, we use the half-life calculations to adjust the time intervals according to the needs of every individual.

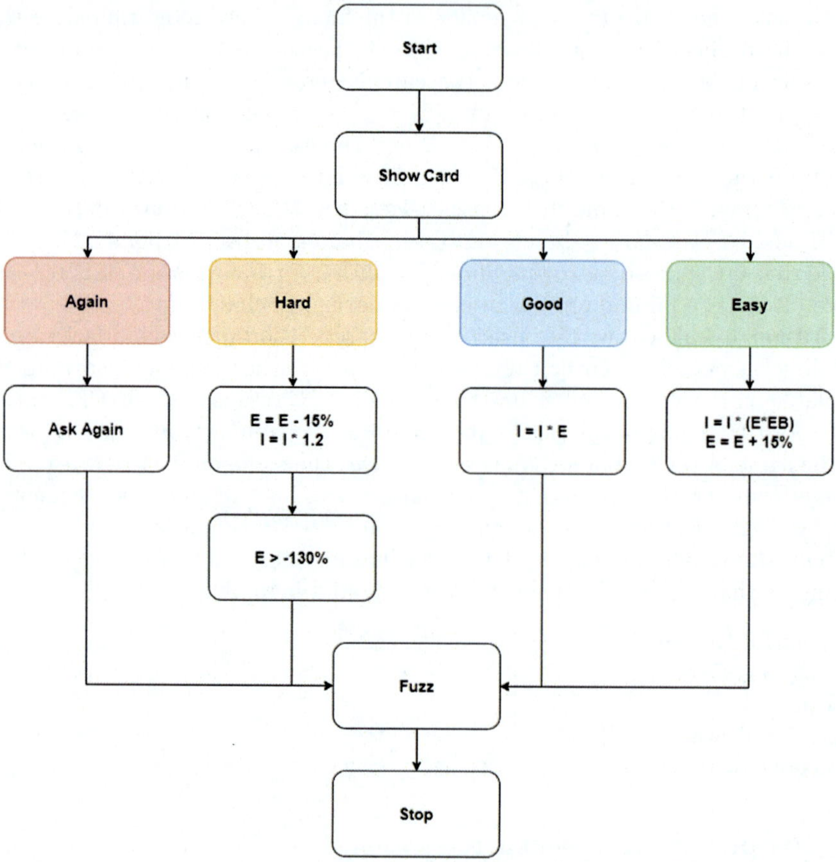

Fig. 3. Proposed architecture.

Another objective we want to achieve is a design that keeps the user engaged. For this we decided to add game-like elements to our interface. Surface level designs that help us retain users. For this, we plan to implement the following elements:

- Leader boards (Local and Global): Here the user will be able to see how they match up to the local and global high scores. This will help them keep motivated.
- Friend's List: Here the user can see their friends. On clicking on their profile, they can see an overview of how their friends perform as compared to themselves.
- Statistical Analysis Page: Here the user can see their performance on different scales across all their subscriptions. Suggestions will also be provided such that the user can improve.
- Progress Bars: Here the user can see a summary of how far they've completed various topics.

- Achievement Awards: For every milestone, users will be awarded an award. This is a positive reinforcement technique that will help students be consistent.

The interval is decided on the basis of how the user responds with their selection of a difficulty with regards to the question they face. Following the Forgetting Curve mapping by Ebbinghaus, we decided to stick to major milestones on the timeline as the interval progression. As shown in the table below, we start with the first repetition after 18 min of going over the card for the first time. This is followed by a repetition after 1 day, which is then followed by a repetition after 6 days, 30 days, 90 days and finally 180 days at which point of time, the interval continues to be 180 days. The basic structure follows that of the proposed infrastructure wherein the user is presented with 3 different difficulties and further repetition-interval calculations are done depending on the difficulty chosen.

Retry: if the user selects this option, first we check if the current interval is equal to the minimum interval which is 18 min. If it is, the interval will continue to remain 18 min till the next repetition. If the repetition is not the minimum 18 min, the interval will be demoted to the previous interval. For example, if the current interval is 30 days, it will be demoted to 6 days.

Difficult: If the user selects this option, there will be no changes in the interval. It will remain the same.

Easy: If the user selects this option, first we check if the current interval is equal to the maximum interval which is 180 days. If it is, the interval will continue to remain 180 d. If the repetition is not the maximum, the interval will be incremented to the next interval. For example, if the current interval is 30 d, it will be incremented to 90 days.

6 Conclusion

To summarize, we propose a new system that aids in memorization, of which the 3 core components are Spaced Repetition, Adaptive Learning and Game-Like Elements. The spaced repetition algorithm will feature a more accurate interval calculation which is based on Ebbinghaus' Forgetting Curve which will make sure that the user is asked to recall the target information just as they're about to forget it, leading to better retention in the future. The adaptive learning algorithm will monitor the user's performance. Based on the data that it collects; it will then adjust the difficulty of the questions asked to the user such that the difficulty doesn't deter them away from the learning process but also doesn't make learning too easy. A balance between challenge and reward will be the key to making sure the system works in the intended manner. For this, the system will feature the implementation of game-like elements such as achievements, badges, awards, etc., which promote a sense of achievement and make users want to achieve their individual goals. Apart from this, we will also add features like leader boards and personal performance statistics that will enable the users to compete with each other, increasing competition and subsequently also increasing the drive to use the system more often.

References

1. Supermemo Algorithm. Accessed Aug 2021. https://www.supermemo.com/en/archives1990-2015/english/ol/sm2
2. Anki. Frequently asked questions. Accessed Aug 2021. https://faqs.ankiweb.net/what-spaced-repetition-algorithm.html
3. Clozemaster. Frequently asked questions, Accessed Aug 2021. https://www.clozemaster.com/faq
4. Shetu, S.F., Rahman, M.M., Ahmed, A., Mahin, M.F., Akib, M.A.U., Saifuzzaman M.: Impactful e-learning framework: a new hybrid form of education. In: Current Research in Behavioral Sciences, p. 100038. Elsevier (2021)
5. Masud, M.A.H., Huang, X.: An e-learning system architecture based on cloud computing. In: System, 2012 - World Academy of Science, Engineering and Technology, vol. 6 (2021)
6. Panicker, R.C., Kumar, A., John, D., Srinivasan, D.: Adaptive learning and analytics in engineering education. In: IEEE International Conference on Teaching, Assessment, and Learning for Engineering (TALE) (2018)
7. Peng, H., Ma, S., Spector, J.M.: Personalized adaptive learning: an emerging pedagogical approach enabled by a smart learning environment. In: Springer Smart Learning Environments (2019)
8. Schimanke, F., Mertens, R., Ribbers, S., Vornberger, O:. Implications of short-term memory research for the design of spaced repetition based mobile learning games. In: IEEE International Symposium on Multimedia (2015)
9. Settles, B., Meeder, B.: A trainable spaced repetition model for language learning. In: 54th Annual Meeting of the Association for Computational Linguistics. Association for Computational Linguistics (2016)
10. Whittman, J.: The Forgetting Curve. California State University, Stanislaus (2018)
11. Yeh, M.K.-C., Toshtzar, A., Guertin, L., Yan, Y.: Using spaced repetition and gamication to enhance K-12 student science literacy with on-demand mobile short reads. In: IEEE Frontiers in Education MISC (FIE) (2016)
12. Zhang, B., Li, Y., Shi, Y., Hou, L.: Recognition and application of learner's cognitive ability for adaptive e-learning. In: IEEE 20th International Conference on Advanced Learning Technologies (ICALT) (2020)

Automation of Supply Chain Management for Healthcare

Hitarth Saiya, Samyak Doshi, Jash Seth(✉), Vishal Badgujar, and Geetanjali Kalme

Department of Information Technology, A.P. Shah Institute of Technology, Thane, India
{hitarthsaiya,samyakdoshi,jashseth,vsbadgujar,
grkalme}@apsit.edu.in

Abstract. Nowadays, the need for systematic deliverance of the healthcare sector is much sought-after. Such, that a simple delay in restocking of certain medicines or supplies could create unrest among the mass, and result in the loss of healthcare facilities for many. Robotic Process Automation (RPA) is a software technology that can create, deploy, and manage bots that emulate certain human actions interacting with digital systems and software. Robotic Process Automation is one of the technologies that can be used to carry out the smooth sailing of the supply chain process. This paper will describe how RPA technology can be used to emulate the various processes in the supply chain management for a healthcare entity, such as checking inventory, reading and comparing vendor quotations, and finalizing a supply vendor.

Keywords: Robotic Process Automation · Supply chain management · Inventory management

1 Introduction

The supply chain of a pharmaceutical entity is the means through which medicines and medical supplies are manufactured and made available to the patients. The traditional Supply Chain of the Healthcare industry involves raw material suppliers who supply raw material to the manufacturers for drug generation, packaging, and supplying to the distributors. The manufacturer maintains an inventory of drugs for smooth operations. The hospital's pharmacy or a retail pharmacy places an order to a distributor for stocking and restocking their inventory, based on the daily requirement.

In a healthcare environment, inventory refers to the medications and all the medical supplies used in its daily operation. A healthcare entity is required to ensure an adequate stock of medicines and medical supplies are available at all times to better serve the needs of their patients. An effective inventory management process should also contribute to the profitability of the entity. There are two goals of an efficient inventory management process. The first is to ensure medications are available whenever the patients need them. Optimal efforts should be made to always keep the medications used in stock and restock them before they go out-of-stock or get outdated and damaged. The second goal is to contribute to the overall profitability of the entity. The focus should be on being

L. Troiano et al. (Eds.): ICDLAIR 2021, LNNS 441, pp. 35–41, 2022.
https://doi.org/10.1007/978-3-030-98531-8_4

of service to the patients while also keeping the purchase costs of the medications at a minimum.

Robotic Process Automation is a business process automation technology that allows an individual or an organization to define a set of instructions for software robots or 'bots' to perform. These bots mimic the way humans interact with software to perform high-volume, repeatable tasks. With the help of RPA, we can automate repetitive tasks such as checking inventory and reordering supplies and decisive processes such as reading and comparing vendor quotations and finalizing a supply vendor. This automation incorporated within the system will make it less prone to errors as well as less time-consuming.

Blue Prism is an RPA Tool that provides a virtual workforce to the organizations. It helps organizations to automate manual, repetitive, and rule-based business processes in an agile and cost-effective manner. It includes drag and drop support to automate the tasks. Some features of Blue Prism include:

1) It is platform-independent so that it can be used on any platform.
2) It contains robust features like load balancing, data encryption, and end-to end auditing. Thus, every change is audited.
3) Blue Prism also supports automation of codes written in Mainframe, Java, Windows applications, and even web-based applications.
4) It supports all major cloud platforms, like Microsoft Azure and Amazon AWS. So, users can manage most of the tasks centrally.

Some advantages of the RPA include:

Cost Savings: RPA helps organizations to save a huge amount of cost as it is typically cheaper than hiring an employee to perform the same set of tasks.
Less Error: It works on standard logic and does not get bored, distracted, or tired. Hence, the probability of making errors reduces to a great extent, which means less re-work and an enhanced reputation for efficiency.
Faster Processing: It works faster than human employees as computer software does not need breaks, food, rest, etc., and can perform repetitive operations tirelessly. With RPA, processing time becomes predictable and consistent, which ensures high-quality customer service across the operations.

The organisation of this paper from hereafter is as follows: Sect. 2 contains the Literature Survey done and Sect. 3 presents the Objectives of the proposed system. Then Sect. 4 describes the Existing System Architecture, methods and the solution to that is shown in the Proposed System Architecture in Sect. 5. Section 6 contains Conclusion and Sect. 7 has Acknowledgment.

2 Literature Review

In this paper [5], the author study aims at the current state of Robotic Process Automation and also recognizes the scientific and industrial literature existing gaps. Here author introduces 54 primary studies analysis which defines the RPA's current state of art. And specifies how these primary studies were selected based on systematic review. Also, the author reviews 14 commercial RPA tools based on 48 functionalities defined. The conclusion of the study is that certain phases except the Analysis phase are covered in majority of the tools and how the shortage of automation in such a phase reflects the technological solutions. Based on this, the author has described some challenges and future directions.

The [6] case study describes how software bots can change our daily life and how the Supply Chain Management process can be impacted once the concept of Robotics comes into the picture. This study describes the different home use cases and day-to-day chores where robotic automation can be used. The author also discusses how RPA has the ability to change the supply chain procedure and eliminate the way of retailing, transporting, and warehousing goods. Process standardization, IT support, the flexibility of the solution, stakeholder expectations, and employee engagement are some of the challenges that might come in the future while implementing RPA in the field of the supply chain.

In this literature [7], the author has given a detailed overview of Robotic Process Automation (RPA). As a result of its increasing popularity, several sectors such as HR, Manufacturing & Retail, Telecom, Healthcare, Insurance, Travel and logistic, Banking & financial services, Government & Infrastructure have started using RPA. RPA tools offer four phases of deployment: i) Assisted Automation, ii) Unassisted Automation, iii) Cognitive RPA, and iv) Autonomous RPA. The author has given a detailed comparison of the three major RPA tools - Automation Anywhere, Blue Prism, and UiPath, out of which UiPath is the most popular tool. Decreasing turn-around-time up to 80%, -30% cost reduction, less time consumption of routine tasks, and proper maintenance of system logs for ISO and auditing purposes are a few of the benefits RPA offers.

In the [7] case study, we reviewed the process model regarding how software testing automation allows us to accomplish the testing process in a faster way than compared to the traditional software testing process. This study differentiates Selenium web Driver and Work Fusion, which are the two methodologies of the software test automation with the new RPA based method. This study recommends combining the test scripts written in different programming languages with the RPA solution for performance improvement, less usage of resources, and making the tests more efficient, reliable, and accurate.

RPA has proven itself to be capable of performing and completing high volume processes which consist of redundant as well as tedious steps. In the research [8], RPA is used to automate the repetitive task of filling out and reading Google forms and input the information filled out in the form, onto a spreadsheet, for educational workshops. Since the software bots follow the exact steps and instructions given to them initially, RPA can effectively perform redundant tasks repetitively without leaving any room for human error.

3 Objectives

1. Maintaining Adequate Stock of medications-
Ensuring there is adequate stock of medications available in the inventory is one of the prime motivations behind the need for efficient management of inventory. The quantity of any given medicine in the inventory should neither be too much that it expires before being sold and nor too little that it's unavailable when needed.

2. Minimizing the Occurrence of Unavailability of medications-
The occurrence of situations where the stock of the required medication isn't available or the medicines in stock are expired should be minimized to decrease the impact on patient care.

3. Reduction of Cost of Maintaining an inventory-
The goal of the system is to decrease the carrying costs of maintaining and updating the inventory while keeping the process smooth and efficient so that it benefits the healthcare organization as well as helps cater to the customers better.

4. Limiting the Use of Resources for Purchasing Tasks
An efficient inventory management system should ensure that minimal resources such as time and human efforts are spent on ordering medications. These resources can be saved or put to better use elsewhere in the system.

5. To Contribute to the Overall Profitability of the Healthcare Organization
The goal is to decrease the total costs to the pharmacy and health care organization by focusing on purchasing products with the lowest price, thus resulting in a profit to the organization. This can be achieved by selecting a vendor that helps keep the acquisition costs as low as reasonably possible.

4 Existing System Architecture

The existing supply chain management framework has mostly relied on the personnel or a group of people handling the system. Also, none of them have a clear idea for the entire product journey as the process flow and teams are isolated. The processes of vendor selection and management of inventory are handled manually. Here, personnel responsible for Vendor Selection reads all the quotations received from the various vendors and compares them based on a single or multiple criteria. The vendor with the best offer is selected, and the personnel updates it on the dashboard.

Another personnel responsible for Inventory Management carries out the tasks of checking the inventory and keeping track of the quantities of the items in stock to determine what medicines or products to order. When the quantityof a product fall below a specified threshold, the personnel manually add it to the list of products to be restocked. Upon receiving the order from the vendor, this personnel again updates the quantity of the item in the database.

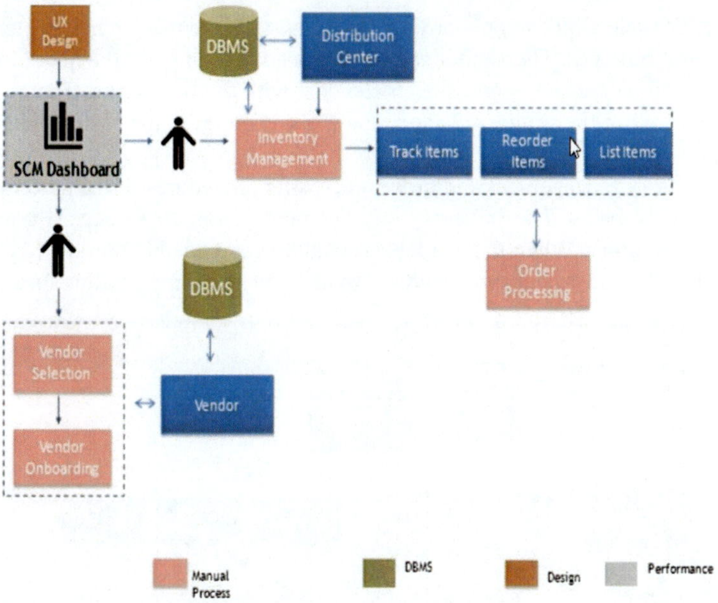

Fig. 1. Existing system architecture.

One of the drawbacks of such a system is that it's more vulnerable to human errors. The team members need to make multiple phone calls or scan through the long email threads to get the desired information which results in ongoing inefficiencies. A minor mistake or negligence by the person handling the system could affect both- the enterprise and its patients/customers. Also not all the incorrectness are being captured and fixed in a timely manner. Another drawback of the current system is that its time-consuming. Certain processes, such as checking inventory, reading and comparing vendor quotations, and finalizing a supply vendor require more attention to details. Thus, the execution of such tasks takes more time when carried out by a human. Instead of spending hours doing repetitive tasks it can be automated for greater accuracy and ability (Fig. 1).

5 Proposed System Architecture

We propose a novel system where software bots will be programmed to carry out processes such as vendor selection, vendor onboarding, inventory management, and order processing which were previously carried out manually. In other words, the repetitive and time-consuming processes of the Supply Chain Management will be automated with the help of Robotic Process Automation (RPA). The incorporation of this automation within the system will make it less prone to errors as well as less time-consuming. Thus, this new proposed system will lead to increased productivity and throughput.

With the help of RPA platform, we create a centralized location to connect all systems which can help the team members to access the required information. So, team members can easily get real-time data and overcomes the disadvantage of the existing system of searching for a particular data from the record book.

The vendor selection and onboarding processes will take place when the tenure current supplier terminates. The vendor selection procedure compares the quotations given by several candidate vendors and selects the deal with the best terms. The onboarding process will finalize the vendor selected, and make changes in the system and database accordingly.

The inventory management and order processing procedures of the system are executed on a daily basis. The inventory management automation keeps a track of the pharmacy's inventory. When the available quantity of any medical product goes below its threshold value, the system will send an email to the supplier regarding the restocking of supplies (Fig. 2).

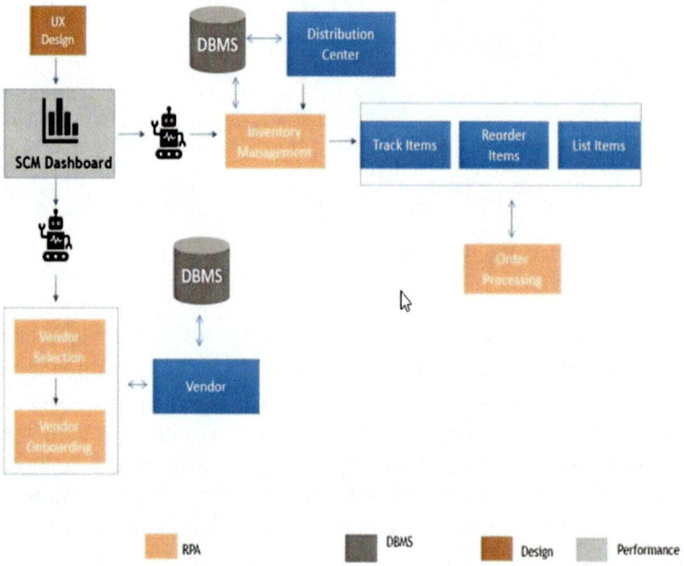

Fig. 2. Proposed system architecture.

Upon receiving an 'Order Confirmed' or 'Order Delivered' email from the supplier, the order processing automation will make the necessary changes in the inventory database. The automation minimizes the time, effort, waste and labour cost and maximizes the productivity and financial improvements.

6 Conclusion

The existing supply chain framework has a lot of shortcomings and we believe these are mainly because of the human management of processes. So, we have proposed a novel supply chain management framework in the healthcare sector, using Robotic Process Automation. By programming BOTs to automate certain processes in a system which were previously handled manually, we can overcome the several shortcomings in the supply chain management system, such as frequent human errors, excess delay, inaccuracies, etc. Automating the key processes, such as Vendor Selection, Vendor Onboarding, Inventory Management and Order Processing with the help of RPA Tool, can help us make the system more reliable and easier to implement.

Acknowledgment. The success of this proposed system would not have been possible without the constant encouragement and advice from a vast number of people. We, the authors of the paper "Automation of Supply Chain Management for Healthcare," would like to express our sincerest regards to Mr. Kiran Deshpande, for their valuable inputs, guidance, encouragement, and constructive criticism throughout our paper. Furthermore, we would like to thank the entire staff of the Computer and Information Technology Department for lending us a helping hand when we most needed it.

References

1. Agostinelli, S.: Synthesis of strategies for robotic process automation, June 2019
2. Aguirre, S., Rodriguez, A.: Automation of a business process using robotic process automation (RPA): A case study (2017)
3. Beetz, R., Riedl, Y.: Robotic process automation: developing a multicriteria evaluation model for the selection of automatable business processes, August 2019
4. Blue Prism: Intelligent Robotic Process Automation – RPA. https://www.blueprism.com/. Accessed 28 July 2021
5. Jain, V.N.: Robotics for Supply Chain and Manufacturing Industries and Future It Holds! (2019). https://doi.org/10.17577/IJERTV8IS030062
6. Rani, A.D.S.G.S.: Recent Trends in Automation-A study of RPA Development Tools (2020).https://doi.org/10.1109/RDCAPE47089.2019
7. Yatskiv, I.V.S.: Improved Method of Software Automation Testing Based on the Robotic Process Automation Technology (2019). https://doi.org/10.1109/ACITT.2019.8780038
8. Sutipitakwong, P.J.S.: The Effectiveness of RPA in Fine-tuning Tedious Tasks (2020).https://doi.org/10.1109/ICEAST50382.2020.9165452

The Detection, Extraction, and Classification of Human Pose in Alzheimer's Patients

David R. Castillo Salazar[1,2](✉), Laura Lanzarini[3], Héctor Fernando Gómez Alvarado[4], and Julio Rafael Cabrera López[1]

[1] Universidad Tecnológica Indoamérica, Ambato, Ecuador
{davidcastillo,juliocabrera}@uti.edu.ec
[2] Facultad de Informática, Universidad Nacional de la Plata, La Plata, Buenos Aires, Argentina
david.castillos@info.unlp.edu.ar
[3] Instituto de Investigación en Informática LIDI (Centro CICPBA), Facultad de Informática, Universidad Nacional de la Plata, La Plata, Buenos Aires, Argentina
laural@lidi.info.unlp.edu.ar
[4] Universidad Técnica de Ambato, Ambato, Ecuador
hf.gomez@uta.edu.ec

Abstract. From a biological perspective, aging is the consequence of the accumulation of extensive molecular and cellular damage over time, which is especially seen in elderly people with neurological diseases such as Alzheimer's who need to move about or who need to move from one side of a room to another, or simply walk and sit because they may feel either anxious or agitated. In this case, the main objective of this research paper is aimed at identifying the estimation of the "Standing" and/or "Sitting" pose in Alzheimer's patients from images obtained from elderly care centers in the canton of Ambato, Ecuador, which is to be used later in an exploratory analysis related to the categories of wandering, nervous, depressed, disoriented, or bored. We worked with a population of 45 people from both sexes, who were diagnosed with Alzheimer's and whose ages ranged between 75 and 89. The methods used were pose detection, feature extraction, and pose classification. As a result, the physical states of "standing" and "sitting" were identified with the usage of algorithms that facilitated the adequate operation of pose estimation. It is concluded that the procedures and data that were obtained provide key input for future research in the health fields related to the behavioral aspects of patients.

Keywords: Classification · Detection · Alzheimer's disease · Extraction · Pose · Neural network · Normalization · Characteristic vector

1 Introduction

The purpose of this research is to understand the estimation of pose from the computational context, for this it is important to reflect that the use of new technologies of artificial intelligence helps to strengthen clinical criteria in the field of medicine, for this the criteria of several authors are supported.

L. Troiano et al. (Eds.): ICDLAIR 2021, LNNS 441, pp. 42–52, 2022.
https://doi.org/10.1007/978-3-030-98531-8_5

Today the use of computational technologies has different approaches when it comes to identifying processes in the clinical context. In [1] it is manifested on cognitive deterioration in early detection and its adequate treatment, this process is related to the patient's gait in RGB images sequences and the estimation of human pose in 3D that allows a correlational analysis with the use of computational algorithms. In [2] it is expressed about patients with mild cognitive impairment who will become Alzheimer's patients. For this, artificial intelligence technologies, in particular the machine learning, applied to neuropsychological measures can lead to a successful automatic classification.

The estimation of human pose currently refers to the processes of capturing information used to analyze joints known as key points: elbows, wrists, among others, as well as body movements obtained by computer vision techniques using an image or a sequence of images to estimate the configuration of the human structure. The 2D pose estimation calculates coordinates of a 2D pose (x, y) for each joint of an RGB image and the 3D pose estimation calculates a 3D pose (x, y, z) coordinates of an RGB image. In [3], the information for the pose of one or several persons in the monitoring of activities is analyzed, as well as the approaches for the estimation, architecture, and procedures of the human pose.

In the research paper elaborated by [4], a regional multi-person pose estimation framework that facilitates body pose in the presence of inaccurate human bounding boxes is presented, which consists of three components: symmetric spatial transformer network, non-maximal parametric pose suppression, and the pose-guided proposal generator. This method achieves an average of 76.7 body recognition points on the human pose dataset: MPII (Multi Person II).

Author [5], proposes a procedure for the location of several individuals and an estimation with the use of key points from the COCO (Common Objects in Context) database in a top-down approach, which focuses on two stages: the position with the box scale with people using the RCNN object detection and location method and the estimation of the key points of each subject contained in a bounding box with the use of ResNet (Residual Neural Network) an average accuracy of 0.649 is achieved in the set of COCO tests.

In the research carried out by [6], the importance of the estimation of the articulated human pose as a fundamental part of computer vision is emphasized. In this work, the important factors in the design of a PRM (Pyramid Residual Module) have been presented to improve the invariances in the DCNN (Deep Convolutional Neural Network) scales, i.e. where the theoretical soundness and efficiency have been demonstrated by experimental analysis.

The present text [7], discusses the basic, simple, and effective methods related to pose estimation, i.e. as identified in categories such as person detection network, human pose estimation network, and flow estimation network and where results are achieved in benchmarks. The best result is shown in the state of the art feature map of 73.7 in the COCO test development division, which has an improvement of 1.6% and 0.7% over the other models.

In the study by [8], a framework for 2D human pose tracking at high frequency is presented–thus maintaining the robustness of state-of-the-art methods by using the Kalman filtering via the OpenPose 2d human pose technique, which is based on deep learning attached to the Lucas-Kanade feature matching method. These facilitate image processing at different frame rates. The results show an improvement in error and higher reconstruction speed as well as fast tracking. The achieved trajectory estimation repeatability was 1 kHz.

In the research carried out by [9], 2D video based fall detection with human pose estimation is proposed using OpenPose to extract the human joint positions. These data are the input of a convolutional neural network and a binary classifier and a SVM (Virtual Support Machine). The results show high values in terms of sensitivity and specificity.

Another related paper is presented by [10], which proposes a fall detection model based on the OpenPose human pose estimation algorithm rooted in computer vision and on the basis of human body key point localization that is combined with SSD (Single Shot MultiBox Detector), which is a MobileNet object identification framework that eliminates non-human key points detected by the OpenPose algorithm. The results show that the method reaches an accuracy rate of 93%.

In this paper [11], the usage of DeeperCut, which refers to a multi-person pose estimation model with a deeper, stronger, and faster scheme, is focused on three stages: the first "deeper" detectors for deep learning of parts of the body, the second "stronger" introduction of image conditioned pairs of the body structure, and the third "faster" image conditioned pairs, which are good candidates for detecting the sections under analysis.

The previously mentioned works show the importance and usefulness of the methods for the detection, extraction, and classification of human pose. In all of them, algorithms are used to provide solutions to different situations that arise in human activities, especially in the elderly.

This document is structured as follows: Sect. 1 identifies relevant research in the specific line of research that was carried out. Section 2 explains the methods and materials used in this work. Section 3 describes the data collection process and the proposed model for the pose classification of Alzheimer's patients. Section 4 presents the results obtained. Finally, Sect. 5 presents the conclusions of the study and future lines of research.

2 Proposed Methodology

The following section presents the materials and techniques used in the detection, extraction, and the classification of pose, i.e. with the usage of the aforementioned algorithms.

A. *Methods*

The methods used focus on three initial aspects: the detection of the pose; second, the feature extraction; and third, the pose classification.

Initially, a reference was made to the detection of the pose of the human structure by identifying the positions in which the parts of the body, arms, legs, head, trunk, etc. are located. For this purpose, the Tensorflow implementation of the DeeperCut algorithm was applied, which takes an input image and detects for each individual the 17 key points as shown in Fig. 1. In [3, 12, 13], information exploration methods for estimating human poses by means of videos and images have been presented by using temporary information gathering techniques and new architectures for processing single and multi-person data.

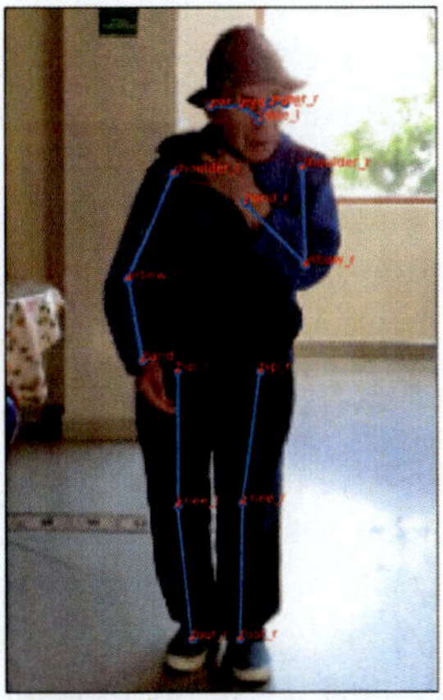

Fig. 1. Pose detection

Feature extraction consists of revealing in the image the position of the elements that are being sought in the human body by generating a numerical representation called feature vector. In [14], the normalization scheme is presented in 2 key stages. The first stage relates to the human body and the second relates to the limbs. Also in [15], a model of Multi-2D pose estimation was identified, i.e. where several types of angles were extracted during the process of motion recognition.

Pose classification is the process of obtaining the spatial configuration of parts of the body using images. The method starts with a reduction of the possible locations of the body structure using color and disparity information of the image. In [16], the steps for the detection and orientation classification are shown. The detected persons are classified as belonging to the following classes: "standing", "sitting on an object", and "sitting on the ground". A deep learning framework was used for this. The results show that the convolutional neural network produces good results.

In Fig. 2, one can see the process that was carried out, which starts with the image input, the pose detection, the extraction of characteristics, and the pose classification–thus obtaining as output the states "standing" and/or "sitting".

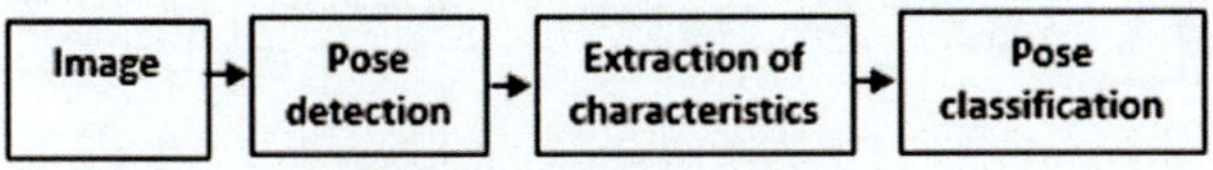

Fig. 2. Processing stages

B. *Materials*

Initially, the process of pose detection in this research focuses on the collation of information from 45 Alzheimer's patients, i.e. where the following factors were taken as a reference: men's ages ranging from 75 to 86 and women's ages ranging from 75 to 89 years– with an initial sample being used for recording data about pose detection, which is indicated in Fig. 1 above. What's more, by using the algorithm called test_pose_detection, key points of the body were identified, which were represented by its coordinates (x,y), and are presented in Table 1.

Next, the process of feature extraction was determined by using the test_features_extraction algorithm, i.e. in order to classify the state of the patient's pose. In addition, it was necessary to build a feature vector from the 17 points detected by the DeeperCut algorithm.

Based on that which is mentioned in the previous paragraph, we first normalized the key points with respect to the center of gravity and body size by following the methodology described in the model's development. In this way, the feature vector will be obtained independently of the patient's size or distance to the camera.

Finally, the classification of the pose was determined by using the test_pose_classification algorithm, which consists of a multilayer neural network that takes the input data from the vector of features and returns it as an output vector of two dimensions corresponding to the probability of "standing" or "sitting", respectively.

The procedure that was carried out required specific equipment consisting of a laptop with a large CPU and GPU capacity for data processing under the Linux Ubuntu operating system. With regard to the software for the development of the algorithms, we used Python 3.5 and tensorflow 1.22 among others, as well as a series of MPII (Human Pose Dataset) data.

The following section discusses the developmental stages of the model for this research paper:

3 Development of the Model

The proposed model has a practical approach that enables us to capture the human body, recognize its key points, and to highlight them so as to later develop its normalization with the data that was obtained, namely to determine the pose classification as shown in Fig. 3.

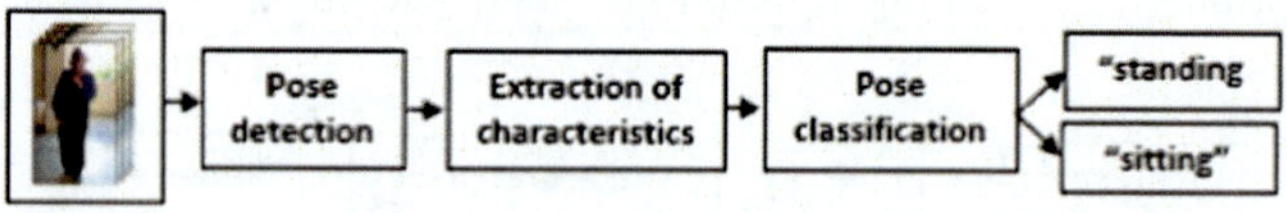

Fig. 3. Classification process

Pose Detection
For this research paper, the detection of the human body was based on a set of images that enabled us to detect the coordinates of key points of the body. For this reason, it

shows the data of elderly individuals while sitting and/or standing. Both Fig. 1 and the results generated in Table 1 in [14,17,18] identify the location of the parts of the human structure from an image, namely 'where a system based on models and body tracking is used and where the methods used are the segmentation techniques and estimation, which is achieved by incorporating important body segments of the human form.

Table 1. Coordinates of the body

Id	Body parts	Coordinates Key points 1	...	Coordinates Key points 45
1	Nose	(91, 42)	...	(86, 28)
2	Eye left	(85, 37)	...	(78, 21)
3	Eye right	(95, 36)	...	(93, 21)
4	Ear left	(74, 37)	...	(66, 28)
5	Ear right	(100, 36)	...	(104, 27)
6	Shoulder left	(61, 61)	...	(45, 57)
7	Shoulder right	(106, 59)	...	(128, 59)
8	Elbow left	(45, 99)	...	(20, 93)
9	Elbow right	(107, 94)	...	(142, 97)
10	Hand left	(50, 128)	...	(29, 123)
11	Hand right	(86, 72)	...	(141, 128)
12	Hip left	(62, 132)	...	(58, 130)
13	Hip right	(91, 132)	...	(105, 132)
14	Knee left	(62, 180)	...	(50, 179)
15	Knee right	(85, 179)	...	(108, 179)
16	Foot left	(67, 228)	...	(54, 226)
17	Foot right	(84, 228)	...	(114, 223)

Feature Extraction

Consequently, once the file with the data of the coordinates of the parts of the human body was made available, we proceeded with the normalization, which initially consisted of obtaining the length of the sections of the human form, which is represented by formula (1).

$$bodylength = length(head + torso + legs) \tag{1}$$

Next, we applied the mathematical procedure of the center of gravity, which is calculated with (2) and (3), that is, where in the length of the body, 17 coordinate values were identified in [19]. It was determined that for each image the coordinates of the singular points were obtained where n is the number of joints related to the length of the body with respect to the center of gravity as well as the individual lengths of the head, torso, and legs. In addition, calculations from the hip to the knee and the other part that covers from the knee to the ankle either on the left or the right side of the body are presented below with the following formulas:

$$Center_x = \sum n \frac{Jn(x)}{bodylength} \tag{2}$$

$$Center_y = \sum n \frac{Jn(y)}{bodylength} \tag{3}$$

The next step is the normalization of the coordinates relative to the body length and the center of gravity. (4) (5).

$$in_x(x) = \frac{jn_x - Center_x}{bodylength} \tag{4}$$

$$in_x(y) = \frac{jn_y - Center_y}{bodylength} \tag{5}$$

Normalized coordinates are values that lie in the range of 0 and 1. Table 2.

Next, the feature vector is constructed containing the values of 34 joint components (17 × 2 coordinates) at points (x,y), as shown in Table 2. In addition, 4 lengths between shoulder to hip and 9 related angles between joints are attached – thereby generating a total of 13 points of body joint angles. Table 3.

The feature vector is completed by totaling the normalized coordinates with the body angles –thus generating 48 new normalized key points as shown in (6).

$$x = \begin{bmatrix} nose,\ left\ eye,\ right\ eye, \ldots, \\ right\ foot,\ shoulder\ length\ from\ left\ to\ left\ hand \ldots, \\ alpha\ left\ shoulder\ left\ elbow\ left\ hand, \ldots, \\ alpha\ head\ neck\ hip \end{bmatrix} \tag{6}$$

Pose Classification

Initially, several data sets with "sitting" and "standing" individuals were reviewed. Although the quest to obtain reliable data was somewhat unsatisfactory, the MPII (Human Pose Dataset), which contains records of images with body joints, helped to satisfy the needs of this research for a significant number of individuals. In fact, 774 "sitting" images of patients and 597 "standing" images of patients were used. From among them, 2 sets of images were elaborated. The first one, which is called SITTING, consists of 221 images where only "seated" individuals appear. Moreover, the second set of images, which is called "STANDING", consists of 222 images of individuals who are "standing".

Neural Network

A multi-layer neural network was used that takes input data from the feature vector XX $\epsilon \hat{R}$ 48 and delivers as output a two dimensional vector $\hat{y}_t = [\hat{y}_1, \hat{y}_2]$, i.e. where

Table 2. Normalized key point coordinates

ID	Body parts	Normalized key point coordinates 1	..	Normalized key point coordinates 45
1	Nose	(0.05643120626470747, –0.3120457337241293)	..	(0.012651786001506555, –0.3546196029648299)
2	Eye left	(0.029483805985028925, –0.3392446368366311)	..	(–0.027133349297486613, –0.3858928293512297)
3	Eye right	(0.07730948107810849, –0.34371417752626904)	..	(0.04543191834328984, –0.38585302532475796)
4	Ear left	(–0.029440411320545538, –0.3367367122244895)	..	(–0.08497209068025353, –0.35618210535834377)
5	Ear right	(0.10107779121459534, –0.34424082488530255)	..	(0.09758258708011706, –0.35733324632002644)
6	Shoulder left	(–0.09288701029039362, –0.21949212791043196)	..	(–0.182701546976272, –0.21568426281272604)
7	Shoulder right	(0.13097706126159028, –0.2277448619176869)	..	(0.21211103985820484, –0.2074626758767918)
8	Elbow left	(–0.1736825098502505, –0.027418904810781932)	..	(–0.30269850247383767, –0.04749254324049822)
9	Elbow right	(0.1391162765337581, –0.05205497999125447)	..	(0.27720215516821956, –0.02656836697481897)
10	Hand left	(–0.14452545407059209, 0.11357004182975552)	..	(–0.2569398676933837, 0.09494439454676407)
11	Hand right	(0.03164168425728021, –0.16132693735644274)	..	(0.27205655922232086, 0.11692130557251099)
12	Hip left	(–0.08772007754405176, 0.13668867507248778)	..	(–0.1223573194606301, 0.12849255377867194)
13	Hip right	(0.055997816660592774, 0.13784563844149814)	..	(0.10230921751844108, 0.13510811290929003)
14	Knee left	(–0.08755499257205138, 0.3765100985656469)	..	(–0.1580687309805238, 0.35925699516495224)
15	Knee right	(0.02943558655143364, 0.37158208681754806)	..	(0.11360608187268055, 0.35728355788472893)
16	Foot left	(–0.0597910770547281, 0.613206255461158)	..	(–0.1396042951064787, 0.5790534548522426)
17	Foot right	(0.02413082289551764, 0.6146171009953245)	..	(0.1415243576040861, 0.5660282835148623)

Table 3. Feature vector Body joints

ID	Body parts	Features vector	.	Features vector
1	length_shoulder_hand_r:	0.1194941748689514	.	0.3298763900504782
2	length_shoulder_hand_l:	0.3370414481752685	.	0.3193767227520757
3	length_hip_foot_r:	0.4778352569634623	.	0.4327008442869881
4	length_hip_foot_l:	0.4773353469918244	.	0.4508908778686439
5	alfa_shoulder_elbow_hand_l:	0.6021090198146652	.	0.9305394603763458
6	alfa_shoulder_elbow_hand_r:	2.4107808453870208	.	0.3812496985869173
7	alfa_hip_shoulder_elbow_l:	2.728907236087172	.	2.348329491870968
8	alfa_hip_shoulder_elbow_r:	2.893012858219295	.	2.4860106811126435
9	alfa_knee_hip_shoulder_l:	0.01381710498943182	.	0.3270997701875063
10	alfa_knee_hip_shoulder_r:	0.0891293778583312	.	0.3609801475942457
11	alfa_foot_knee_hip_l:	0.0138171049894318	.	0.3270997701875063
12	alfa_foot_knee_hip_r:	0.0913326534036146	.	0.0821517915646649
13	alfa_knee_l_hip_l_knee_r_hip_r:	0.1138447122018487	.	0.2043375969676733

$\hat{y}_1$ and $\hat{y}_2$ that correspond to the probability of sitting or not sitting and the schemata of the neural network is shown in Fig. 4. The number of neurons in the first, second, third, and fourth layers is 64, 32, 16 and 2, respectively. The activation function was applied in the input layer. In the hidden layers, the Relu function was used. Furthermore, in the output layer, the Softmax function is identified in [18] and [19]. Neural network architectures were used in order to show the reliability of specific models for the diagnosis of Alzheimer's disease. In addition, an image dataset with the data normalization procedure and training architecture of the DNN artificial neural network (Deep Neural Network) and MLP is described in [20]: DNN (Deep Neural Network) and MLP (Multi-Layered Perceptron).

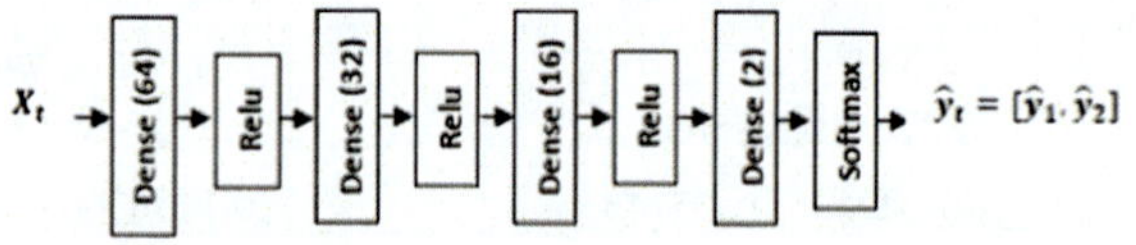

Fig. 4. Neural network architecture

The neural network allows us to identify an output 1, which is represented by "standing" and a zero—and subsequently represents "sitting" as follows: [0, 1] stopped; [1 0] sitting, e.g. as seen in terms of the identification of the images and the usage of a tracker that assigned with the letters S=Sitting and P=Stopped. This identifier is assigned to each individual regarding the emotional state that is being represented in the frame. In [20], some basic human-centered behaviors (walking, walking up, walking down, sitting, standing, and lying down) were identified.

The following section presents the results of the research that was carried out.

4 Results

With the usage of algorithms for detection and extraction, as well as for identifying the characteristics and classification of the pose of Alzheimer's patients of an average age, data from the pose of the body was registered for each one of the algorithms previously mentioned. This process allowed us to identify with the labels (P = "Standing") and (S = "Sitting") patients with the neurological disease, thereby fulfilling an important stage of the proposed objective.

The following section presents the conclusions of the research that was carried out.

5 Conclusions

Post detection is a fundamental process because it allows one to locate the key points of the human body in coordinates (x,y) within a 2D context.

Feature extraction is likewise relevant because it enables us to normalize certain key points of the body. Moreover, we are subsequenly able to obtain the coordinates of the normalized points – thereby defining a complete feature vector for its subsequent development.

Pose classification is based on a neural network and confirms that the above procedures are reliable. Moreover, the results obtained confirm the statuses of "sitting" or "standing" according to the input data that was evinced at the start of the process.

Finally, the states of "standing" and "sitting" that are based on an input image were also confirmed. This input data may thus serve as stimulus for future work and help to determine other similar characteristics of patients using a probable mathematical model – thereby generating potential behavioral alerts exclusively in patients with neurological diseases, which in this case is Alzheimer's disease.

References

1. Kondragunta, J., Hirtz, G.: Gait parameter estimation of elderly people using 3D human pose estimation in early detection of dementia. In: pp. 5798–5801 (2020)
2. Battista, P., Salvatore, C., Berlingeri, M., Cerasa, A., Castiglioni, I.: Artificial intelligence and neuropsychological measures: the case of Alzheimer's disease. Neurosci. Biobehav. Rev. **114**, 211–228 (2020)
3. Munea, T.L., Jembre, Y.Z., Weldegebriel, H.T.: The progress of human pose estimation : a survey and taxonomy of models applied in 2D human pose estimation. IEEE Access. **20**, 133330–133348 (2020)
4. Fang, H.S., Xie, S., Tai, Y.W., Lu, C.: RMPE: regional multi-person pose estimation. In: Proceedings of IEEE International Conference on Computer Vision, vol. 2017-October, pp. 2353–2362 (2017)
5. Papandreou, G., et al.: Towards accurate multi-person pose estimation in the wild. In: Proceedings of 30th IEEE Conference on Computer Vision Pattern Recognition, CVPR 2017, vol. 2017-January, pp. 3711–3719 (2017)

6. Yang, W., Li, S., Ouyang, W., Li, H., Wang, X.: Learning feature pyramids for human pose estimation. In: Proceedings of International Conference on Computer Vision, vol. 2017-October, pp. 1290–1299 (2017)
7. Xiao, B., Wu, H., Wei, Y.: Simple baselines for human pose estimation and tracking. In: Ferrari, V., Hebert, M., Sminchisescu, C., Weiss, Y. (eds.) ECCV 2018. LNCS, vol. 11210, pp. 472–487. Springer, Cham (2018). https://doi.org/10.1007/978-3-030-01231-1_29
8. Landolfi, L., Tripicchio, P., Filippeschi, A., Avizzano, C.A.: Fast and fluid human pose tracking. In: 2019 IEEE International Conference on Real-time Computing and Robotics, pp. 24–29 (2019)
9. Huang, Z., Liu, Y., Fang, Y., Horn, B.K.P.: Video-based fall detection for seniors with human pose estimation. In: 4th IEEE International Conference on University Village 2018, UV 2018, pp. 1–4 (2019)
10. Sun, G., Wang, Z.: Fall detection algorithm for the elderly based on human posture estimation. In: pp. 172–176 (2020)
11. Insafutdinov, E., Pishchulin, L., Andres, B., Andriluka, M., Schiele, B.: DeeperCut: a deeper, stronger, and faster multi-person pose estimation model. In: Leibe, B., Matas, J., Sebe, N., Welling, M. (eds.) ECCV 2016. LNCS, vol. 9910, pp. 34–50. Springer, Cham (2016). https://doi.org/10.1007/978-3-319-46466-4_3
12. Cao, Z.: Realtime multi-person 2d pose estimation using part affinity fields. In: pp. 1302–1310 (2017)
13. Li, Y., Li, K., Wang, X., Yi, R., Xu, D.: Exploring temporal consistency for human pose estimation in videos. Pattern Recogn. **103**, 107258 (2020)
14. Sun, K., Lan, C., Xing, J., Zeng, W., Liu, D., Wang, J.: Human pose estimation using global and local normalization. In: Proceedings of IEEE International Conference on Computer Vision, vol. 2017-October, pp. 5600–5608 (2017)
15. Cao, G., Pu, Y., Li, Y.: Human motion capture using a multi-2D pose estimation model. In: 2019 11th International Conference on Intelligence Human-Machine System and Cybernetics, vol. 1, pp. 64–67 (2019)
16. Atvar, A.: Insan Duru ¸ s ve Yönelimlerinin Derin Ö ˘ grenme ile Sınıflandırılması Classification of Human Poses and Orientations with Deep Learning, pp. 2–5 (2018)
17. Jalal, A., Nadeem, A., Bobasu, S.: Human body parts estimation and detection for physical sports movements. In: 2019 2nd International Conference on Communication, Computing and Digital Systems, C-CODE, pp. 104–109 (2019)
18. Takano, W., Lee, H.: Action description from 2D human postures in care facilities. IEEE Robot. Autom. Lett. **5**(2), 774–781 (2020)
19. Pismenskova, M., Balabaeva, O., Voronin, V., Fedosov, V.: Classification of a two-dimensional pose using a human skeleton. In: vol. 05016, pp. 10–13 (2017)
20. Ghosh, N.S., Majumdar, R., Giri, B., Ghosh, A.: Detection of human activity by widget. In: ICRITO 2020 - IEEE 8th International Conference on Reliability Infocom Technologies and Optimization (Trends Future Directions), pp. 1330–1334 (2020)

Simulating Using Deep Learning The World Trade Forecasting of Export-Import Exchange Rate Convergence Factor During COVID-19

Effat Ara Easmin Lucky(✉), Md. Mahadi Hasan Sany, Mumenunnesa Keya, Md. Moshiur Rahaman, Umme Habiba Happy, Sharun Akter Khushbu, and Md. Arid Hasan

Department of Computer Science and Engineering, Daffodil International University, Dhaka 1209, Bangladesh
{effat15-10793,mahadi15-11173,mumenunnessa15-10100, moshiur15-11112,umme15-10916,sharun.cse, arid.cse0325.c}@diu.edu.bd

Abstract. By trade we usually mean the exchange of goods between states and countries. International trade acts as a barometer of the economic prosperity index and every country is overly dependent on resources, so international trade is essential. Trade is significant to the global health crisis, saving lives and livelihoods. By collecting the dataset called "Effects of COVID-19 on trade" from the state website NZ Tatauranga Aotearoa, we have developed a sustainable prediction process on the effects of COVID-19 in world trade using a deep learning model. In the research, we have given a 180-day trade forecast where the ups and downs of daily imports and exports have been accurately predicted in the Covid-19 period. In order to fulfill this prediction, we have taken data from 1st January 2015 to 30th May 2021 for all countries, all commodities, and all transport systems and have recovered what the world trade situation will be in the next 180 days during the Covid-19 period. The deep learning method has received equal attention from both investors and researchers in the field of in-depth observation. This study predicts global trade using the Long-Short Term Memory. Time series analysis can be useful to see how a given asset, security, or economy changes over time. Time series analysis plays an important role in past analysis to get different predictions of the future and it can be observed that some factors affect a particular variable from period to period. Through the time series it is possible to observe how various economic changes or trade effects change over time. By reviewing these changes, one can be aware of the steps to be taken in the future and a country can be more careful in terms of imports and exports accordingly. From our time series analysis, it can be said that the LSTM model has given a very gracious thought of the future world import and export situation in terms of trade.

Keywords: Deep learning · Export value · Import value · Forecasting · Time series analysis · Long-Short Term Memory

L. Troiano et al. (Eds.): ICDLAIR 2021, LNNS 441, pp. 53–62, 2022.
https://doi.org/10.1007/978-3-030-98531-8_6

1 Introduction

The world has become a global village and trade plays the main role between the countries as an interaction channel. Trade is a process of exchanging products and services between states or countries and moreover, it can see in between different companies, or any other substances. Trade plays an awfully important part in setting up a higher global economy. We can talk about China's trade as an example of world trade's importance. By addressing "World Factory" China is known in numerous nation's views to the world currently, and the trademark "Made in China" shows up in nearly every corner of the world [2]. Theories about worldwide trade expanding total efficiency at the nation's level are about as ancient as economics [3]. By the word 'world trade' or 'foreign trade', means that it is a process in which required goods or services are received and domestic products are sold in the global market [4]. Import and export are the two most important words in world trade, where import means purchasing foreign products or services and bringing them to one's home country and export means selling of products from home country to a foreign country. The relationship between foreign trade and economic growth is still a topic of debate from ancient times till now [5]. As a major factor of openness, worldwide trade has made a progressively significant contribution to economic growth [6]. This relationship (between foreign trade and economic growth) is based on two approaches-one is growth theory based on import and another one is growth theory based on export [7]. To capture the perfect view of the relationship between foreign trade and economic growth, time series analysis is always the best way [8]. The time series analysis method is utilized to analyze and understand past data and predict the future, generally, time-series data sets are annual data of different periods. When a set of observations is organized chronologically, then it is called a time series, time series analysis forms one of the most significant tools of the economist [9]. Long-term time series have various tricks or strategies like recursive, direct, and DirRec which are examined and compared [10]. In our paper, we collect the world trade data for the period January 01, 2015–May 30, 2021. The main purpose of the investigation is to compare the import and export data for all countries for the period January 1st, 2015–May 30, 2021, and based on the result the system predicts the future possibility of world trade. That's why we apply the time series model LSTM to analyze the data and prepare future predictions for the next 180 days of 2021.

2 Literature Review

Trade is an essential financial idea which counts the buying and selling of goods or projects and services, through paying payment by a buyer to a seller. From different papers, in this part, here reviewed the same categorical purposes. From the study, the findings are- a lot of variation in research with trade, many of them include here like, Avi et al. [11] investigate the global components of the economics of AI. In the discussion, trade theory which analyses the roles of scales, rivalry, and information creation, and information diffusion essential to near advantage comes up. Reuven et al. [12] describes a question in the work, the question is "Does a currency union affect trade?". The description of it's given with time series and assesses the impact of currency unions on trade abusing time series variety and the data amount of trade was huge between 1970 and 1990. Both time series and cross-sectional variety of currency union rates had been utilized. Shahbaz et al. [13] explored how much vitality request is influenced by urbanization, financial improvement, and trade openness in essentialness request work within the case of Malaysia. They collected data from world development indicators and the data period was between 1971 to 2011, and by using the Interpolation method they convert annual data series into quarterly frequency. Jungho et al. [14] analyze the dynamic relationships between energy consumption, income growth, trade, and CO_2 outflows for G-20 economies in a system of CVAR (cointegrated vector autoregression). The result shows that economic growth and trade have an inclinable impact on natural quality for the created G-20 portion countries. The data was collected from World Development Indicators (WDI). Shuanglin et al. [15] analyzed the trade and economic growth relationship and the data was China's national data from 1952 to 1997 and the output of their work was relatively positive with the growth rate of trade and growth rate of per capita GDP. Utku et al. [16] analyzed the need for imports and exports for Turkey concerning the EU. They applied time series analysis for the period 1963 to 2002 and used annual data. Their experimental results recommend that the Turkish import demand from the EU is obliged by its external obligation. Future prediction of trade is essential to know because it helps us to handle any circumstances. Rana et al. [17] examined the pattern of trade and estimated the future trend of Pakistan by using time series analysis for the period 1960–2010. The outcomes show that the export structure has changed essentially since the 1960s by a keen drop in the share of essential products and increment of industrial products. Exploring relationships between trade, FDI, and growth rate of per capita GDP is very important to the analysis. Mohammed et al. [18] investigated the relationship of the trade, foreign direct investment, and the growth rate of per capita GDP of Bangladesh from 1973 to 2014 with the help of time series analysis. The data was collected from the World Bank online database, their VECM model (Vector Error Correction Model) shows all of the variables of their paper have a long-term relationship. For the validity check of the VECM model, they applied several post estimation diagnostic tests and as result, they found that the residuals of the regressions have typical conveyance and don't appear to have any auto-correlation. Melina et al. [19] also applied time series analysis for the period 1960–2002 and analyzed the relationship of trade, foreign direct investment, and economic growth of Greece and as a result, they found a long-run balanced relationship between the mentioned variables. They collected annual data from the International Monetary Fund for the period 1960–2002 and applied the VAR (Vector Auto Regressive model) for determining the relationship between trade, foreign direct investment, and economic growth. Moreover, for the period of 1960–2005 in Turkey, the relationship between energy consumption, CO_2 emissions, income, and foreign trade was analyzed by Halicioglu et al. [20]. They collected data from IMF (International Financial Statistics), and World Bank (WB) which is World development indicators, and Annual Statistics of Turkish Statistical Institute (TSI), and applied different models including the ARDL model. The experimental outcomes of their paper proposed that in Turkey, income is the most essential variable for the explanation of carbon emissions which is taken after by energy consumption and foreign trade.

3 Methodology

The workflow of our work is given below (Fig. 1)

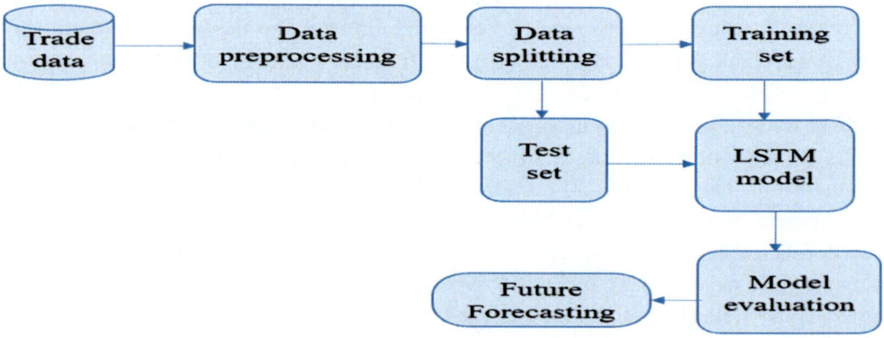

Fig. 1. FlowChart of world trade time series predictor

3.1 Data Collection

Data collection is a process of gathering or collecting data and then estimating and analyzing them for a proper overview of research utilized standard endorsed procedure. Data collection is the very first and most important part of any research work. There are many different processing of collecting data based on the research work. We can collect data from different social networking media and websites [21]. And thus, online data collection is becoming a progressively well-known research strategy [22]. For the work, we collect trade data from the States NZ Tatauranga Aotearoa website. There are import and export data from January 1st, 2015–May 30, 2021 for all countries, all commodities, and all transport systems. Then we analyze these data and predict the future for the next 180 days. The daily basis global import & export values are shown below in Fig. 2.

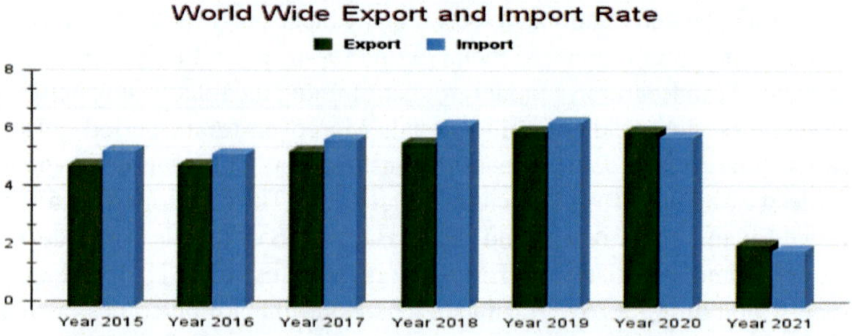

Fig. 2. Yearly import & export data analysis on time series data.

3.2 Long-Short Term Memory

Long Short-Term Memory is known as a type of artificial recurrent neural network and it is utilized in the deep learning field for providing long-term association. LSTM works better than traditional recurrent neural networks (RNN) [23]. It was influenced by the analysis of error flow in the existing RNN. LSTM has hidden layers and in hidden layers, there is a set of blocks which are known as memory blocks, these blocks have differentiable memory chips in advanced computers [24]. These memory chips have one or more repetitively involved memory cells and also have three multiplicative units including input, output, and input gates. And these units provide continuous analogs of reading, write and reset operations for the cells. Net is used for interacting with the cells through the gates. We use adam optimizer because it attaches the most excellent properties AdaGrad and RMSProp algorithms to supply an optimization algorithm that can manage inadequate slopes on riotous issues. The adam optimizer is much faster than other optimization methods in neural networks [25]. We also use Mean Square Error for minimizing loss (Table 1).

Table 1. Table (a) & (b) represent the LSTM model summary

Layer(type)	Output shape	Param
lstm_6 (LSTM)	(1, 1, 4)	96
lstm_7 (LSTM)	(1, !)	24
dense_3(Dense)	(1, !)	2

Table-(a)

Total params	Trainable Params	Non-trainable params
122	122	0

Table-(b)

3.3 LSTM Model Estimation with the Parameters

The parameters of the model are set in the LSTM model very easily with the appropriate input in real time transmission. We work with 60% and 30% data for training and testing and where batch size and epoch are 1 and 100 respectively (Table 2).

3.4 LSTM Model Analysis

For a strong representation of future predictions, a comparative study is very essential. From the dataset of world trade, significant with LSTM where the Long and Short-Term groupings and it can be extricated the procedure. We trained the export-import trade data from January 1st, 2015 to May 30, 2021, per day. World trade study prediction is trained by LSTM which gives the actual prediction [26] (Table 3).

Table 2. LSTM transcendence order

LSTM transcendence order			
Training size	Testing size	Batch size	Epochs
70%	30%	1	100

Table 3. Model loss on gobal exchange rate.

	Export		Import	
	Highest loss	Lowest loss	Highest loss	Lowest loss
All over world	0.0507	0.0378	0.2652	0.0081

4 Results and Discussion

Figure 3 is visually representing the training testing values globally for import. The blue, yellow, green, and red waves give the train (true & predicted value), test (true & predicted value) consequently. We can see that the training true value and testing predict values are relatively close. The training testing values globally for Imports start from 2015 and end in 2021.

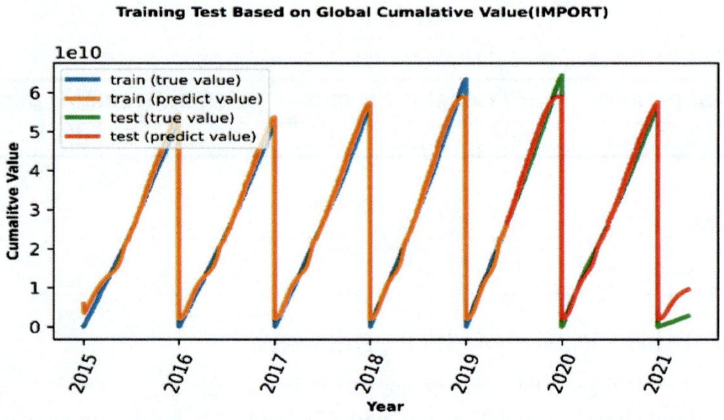

Fig. 3. Yearly importing true & predicted cumulative values using LSTM.

Figure 4 has plotted the training testing values globally for export. In this graph, the blue wave indicates the train (true value), the yellow wave indicates train (predict value), the green wave indicates test (true value), and the red wave indicates test (predict value). That train (true value) is relatively close to the test (true value), and the train (predict value) is relatively close to the test (predict value). The training testing values globally for exports start from 2015 and end in 2021.

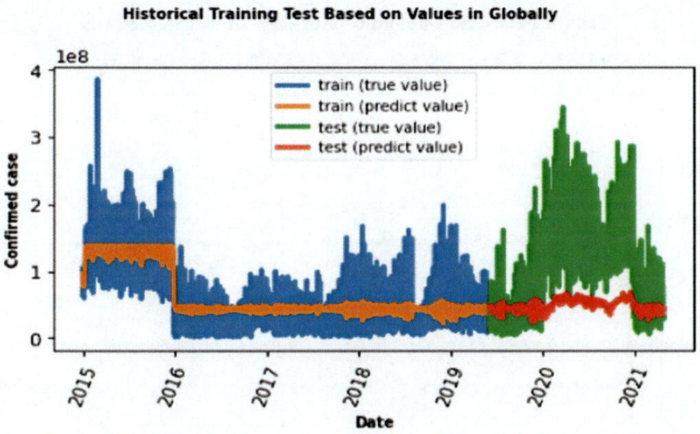

Fig. 4. Yearly exporting true & predicted cumulative values using LSTM

In Fig. 5, we can see the future prediction on per day Cumulative values of global import. In this graph, it has seen that the value 2.56 is constant with the date. The beginning date of the prediction is 2021-05 and the ending date is 2021-11.

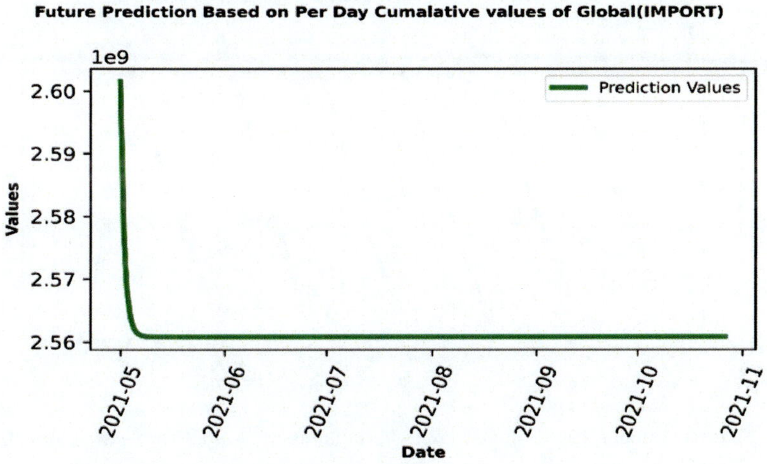

Fig. 5. Future trajectory based on per day import cumulative values.

Figure 6 visualizes the future prediction on per day Cumulative values of global export. Future prediction for each day is shown in the given graph. It is also seen that the value 4.94 is constant for a long time. The beginning date of the prediction is 2021–05 and the ending date is 2021–11.

Figure 7 has plotted the future prediction of global imports for the next 180 days. It shows that the result will increase constantly in the next 180 days (from June 2021 to November 2021). By comparing the previous data and our result, we can say that our model estimation is so much true.

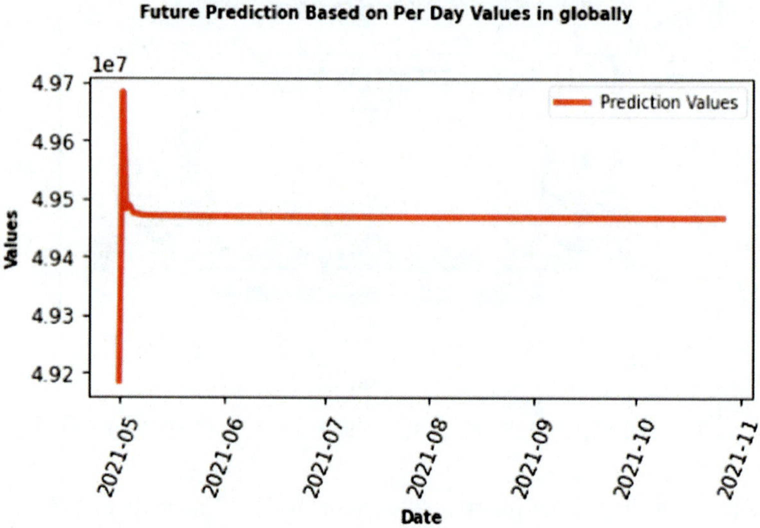

Fig. 6. Future trajectory based on per day export cumulative values.

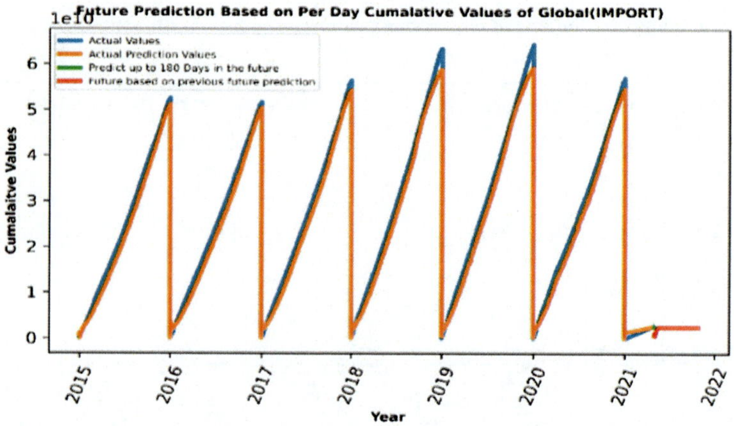

Fig. 7. Future prediction based on per day import cumulative values for next six months using LSTM.

Figure 8 has plotted the future prediction of global exports for the next 180 days. It shows that the result will increase constantly in the next 180 days. We can say that our model estimation is so true after comparing the previous data and our result.

We wanted to predict future for world trade. And we are successful on our work. Our main focus/choice is to predict future by using time series analysis and we have done it smoothly by applying LSTM model. By comparing our results with more relatable works result [16, 17], we can say that in both result, future prediction have done with LSTM and time series analysis.

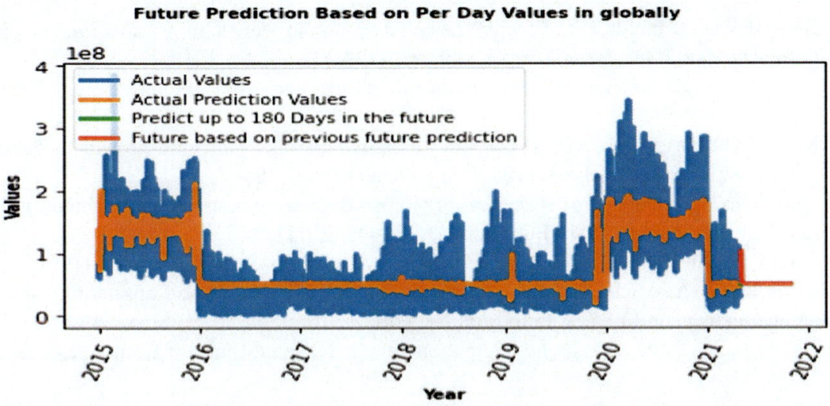

Fig. 8. Future prediction based on per day export cumulative values for next six months using LSTM

5 Conclusion and Future Work

This research provides an idea of what the future holds for imports and exports in world trade during the Corona period with the LSTM model. The dataset, titled "Effects of COVID-19 on trade", was collected from the state website NZ Tatauranga Aotearoa to provide time-series analysis and is the original source of the data. The time series analysis has been studied based on export and import data from 2015, January 1 to 2021, May 30, for all countries, all commodities, and all transport systems. In the methodology, we have tried to predict what might happen in the next 180 days in world trade through various graphical visualizations which have used cumulative value in global trade in terms of imports and exports. The 180-day forecast shows that in the Covid-19 period the cumulative value of imports and exports is likely to remain unchanged from May 30, 2021 to November 2021.

We have some limitations in this work as we have only worked and predicted cumulative value but what the global economic situation might be in terms of trade was not discussed during this coronal period or no graphical representation was shown. In future, we'll try to create comparative study between the time series analysis of LSTM & ARIMA, SRIMA model.

References

1. Serrano, M.A., Bogu, M.: Topology of the world trade web. Phys. Rev. E **68**, 015101(2015)
2. Yu, L., Wang, S., Lai, K.K.: Forecasting China's foreign trade volume with a kernel-based hybrid econometric ai ensemble learning approach. J. Syst. Sci. Complex. **21**, 1–19 (2008). https://doi.org/10.1007/s11424-008-9062-5
3. Alcala, F., Antonio Ciccone, U.D.M., Fabra, U.P.: Trade and productivity. Q. J. Econ. **119**, 613–646 (2004)
4. Nosov, V., Tcypin, A., Mineeva, L., Pshentsova, A., Mukhina, I.: Time Series Research of Foreign Trade Operations of Post-Soviet Countries. Atlantis Press (2019)

5. Berg, H.V.D., Schmidt, J.R.: Foreign trade and economic growth: time series evidence from Latin America. J. Int. Trade Econ. **3**, 249–268 (1994)
6. Velnampy, T.: Export, import and economic growth: evidence from Sri Lanka. J. Econ. Sustain. Dev. **4**, 147–155 (2014)
7. Kilc, N.O., Beser, M.: Relationship of foreign trade and economic growth in eurasian economy: panel data analysis. Int. J. Econ. Finance **9**, 1–7 (2017)
8. Bakari, S.: The relationship between export, import domestic investment and economic growth in Egypt: empirical analysis. Euro Econ. **36**, 34–43 (2017)
9. Parzen, E.: An approach to time series analysis. Ann. Math. Stat. **32**, 951–989 (1961)
10. Grigorievkiy, A., Miche, Y., Vental, A.M., Everin, E.S., Lendasse, A.: Long-term time series prediction using OP-ELEM. In: Neural Networks (2014)
11. Goldfarb, A., Trefle, D.: AI and international trade. National Burea of Economics Research (2018)
12. Glick, R., Rose, A.K.: Does a currency union affect trade? The time series evidence. National Burea of Economic Research (2001)
13. Muhammad, S., Loganathan, L., Sbia, R., Talat, A.: The effect of urbanization, affluence and trade openness on energy consumption: a time series analysis in Malaysia. In: Renewable and Sustainable Energy Reviews (2015)
14. Baek, J., Kim, H.S.: Trade liberalization, economic growth, energy consumption, and the environment: time series evidence from G-20 economics. J. East Asian Econ. Integr. (2011)
15. Lin, S.: Foreign trade and China's economic development: a time series analysis. J. Econ. Dev. **25**, 145–154 (2000)
16. Utkulub, U., Seymen, D.: Trade and competitiveness between Turkey and the EU. Turkish Economic Association (2004)
17. Anwar, R.M.A., Nasreen, S., Chaudhary, M.A.: Patterns of trade and its future prospects in Pakistan: a time series analysis. Br. J. Arts Soc. Sci. **13**(2), 228–241 (2013)
18. Hussain, M.E., Haque, M.: Foreign direct investment, trade and economic growth: an empirical analysis of Bangladesh. In: Economic Development in Southeast Asia (2016)
19. Dritsaki, M., Dritsaki, C., Adamopoulos, A.: A causal relationship between trade, foreign direct investment and economic growth in Greece. Am. J. Appl. Sci. **1**(3), 230–235 (2004)
20. Halicioglu, F.: An econometric study of CO_2 emissions, energy consumption, income and foreign trade in Turkey. In: 31st IAEE Annual International Conference (2009)
21. Durresi, L.A.A.: Techniques for collecting data in social networks. In: 2014 International Conference on Network-Based Information System (2014)
22. Granello, D.H., Wheaton, J.E.: Online data collection: strategies for research. J. Couns. Dev. **82**, 387–393 (2004)
23. Gers, F.A., Schraudolph, N.N., Schmidhuber, J.U.: Learning precise timing with LSTM recurrent networks. J. Mach. Learn. Res. **3**, 115–143 (2002)
24. Graves, A., Fernández, S., Schmidhuber, J.: Bidirectional LSTM networks for improved phoneme classification and recognition. In: Duch, W., Kacprzyk, J., Oja, E., Zadrożny, S. (eds.) ICANN 2005. LNCS, vol. 3697, pp. 799–804. Springer, Heidelberg (2005). https://doi.org/10.1007/11550907_126
25. Bock, S., Goppold, J., Weib, M.: An improvement of the convergence proof of the ADAM-optimizer. In: Conference Paper at OTH CLUSTERKONFERENZ (2018)
26. Saqib, M., Sha, Y., Wang, M.D.: Early prediction of sepsis in EMR records using traditional ML techniques and deep learning LSTM networks. In: 40th Annual International Conference of the IEEE Engineering in Medicine and Biology Society (EMBC) (2018)

Leveraging Free-Form Text in Maintenance Logs Through BERT Transfer Learning

Syed Meesam Raza Naqvi[1,2](✉), Christophe Varnier[1], Jean-Marc Nicod[1], Noureddine Zerhouni[1], and Mohammad Ghufran[2]

[1] FEMTO-ST Institute, Université de Bourgogne Franche-Comté/CNRS/ENSMM, Besançon, France
{syedmeesam.naqvi,christophe.varnier,jm.nicod, noureddine.zerhouni}@femto-st.fr

[2] Fives CortX, Lyon, France
{meesam.naqvi,mohammad.ghufran}@fivesgroup.com

Abstract. Across various industries, maintenance entries recorded over time contain decades of experience and health records of different assets. Maintenance logs are usually free-form text entries recorded by maintenance operators. These records are also highly unstructured and imbalanced. Because of these reasons, this huge source of knowledge is usually underutilized and does not contribute to the development of tools to help improve maintenance processes. In the last few years, due to recent advancements in the field of Natural Language Processing (NLP) and increased focus on industry 4.0, there is a need to revisit this problem. This study explores the use of state-of-the-art NLP methods on free-form maintenance text to leverage this data. More specifically, the purpose of the study is to estimate the problem category of maintenance log to see how well recent NLP models adapt to free-form maintenance text. Findings of this study indicate that CamemBERT with the fine-tuning approach outperforms the classical NLP approaches. Class imbalance problem is also addressed through data augmentation using deep contextualized embedding showing further performance improvement. Model accuracy and Matthews correlation coefficient (MCC) are used as performance metrics to give a better understanding of results with imbalanced classes.

Keywords: Industry 4.0 (I4.0) · NLP · BERT · Maintenance logs · Classification

1 Introduction

As we are moving forward into the era of Industry 4.0 (I4.0), data has proved to be the key player in the industrial revolution. Data-driven Artificial Intelligence (AI)

L. Troiano et al. (Eds.): ICDLAIR 2021, LNNS 441, pp. 63–75, 2022.
https://doi.org/10.1007/978-3-030-98531-8_7

powered solutions are transforming industrial processes. Due to these advancements in the field of AI and big data analytics, data is now called 'the new oil' [6]. Across various industries, documents recorded over time contain decades of knowledge obtained through experience. It also contains valuable health records and maintenance histories. The main problem is that the information in these documents is technical and is mostly free-form (unstructured) text. Natural language processing (NLP) is a branch of computer science that deals with the application of computational methods for the automatic analysis and representation of human language [11]. Contrary to regular languages, in maintenance logs, both vocabulary and the document structure are significantly different from data used in training general NLP pipelines. As a result, this information cannot be readily processed using these pipelines. To address and formalize this issue, General Electric (GE) Digital, in collaboration with researchers from the National Institute of Standards and Technology (NIST) and the University of Western Australia, recently published a study. In this study, they proposed Technical Language Processing (TLP) as a sub-domain of AI. TLP can be considered as a way to adopt NLP pipelines for industrial use cases to develop reproducible and scalable industrial solutions [3]. Across industries, Maintenance Work Orders (MWO) are among common sources of technical language data. MWO planning, scheduling, and execution generate a significant amount of data over the life cycle of assets. These data, collected over the years, contain information about asset performance, maintenance policies, and failure patterns. If efficiently utilized, this information can be converted into actionable intelligence. However, the information in maintenance records is usually underutilized because 1. vocabulary and sentence structures vary significantly from natural language texts, 2. Information is not structured and not uniformly presented, and 3. Lack of good quality data. Recent trends in NLP have minimized the dependency on these classical preprocessing pipelines. In this study, we will explore the capability of state-of-the-art techniques in NLP on unstructured maintenance text. To process text using a Machine Learning (ML) algorithm, the first step is to convert it into numerical representation through vectorization. One of the widely used classical approaches for vectorization of textual information is Term Frequency-Inverse Document Frequency (TF-IDF). Given a set of documents, the idea behind TF-IDF is to convert each document into a vector by giving more importance to the terms that are unique to a given document compared to common terms in all available documents. In TF-IDF, frequency of each term is multiplied by a weight that depends on the rarity of that term in the common vocabulary of all documents [7]. TF-IDF is one of the most widely used techniques in the field of NLP [1].

In 2013, there was a shift in NLP research with the introduction of non-contextualized word embedding, a vector representation of words. Techniques like Word2Vec by Google [12], GloVe by Stanford [13] and FastText by Facebook [2] transformed classical preprocessing pipelines. FastText can divide a word into subforms, thus automatically handling misspelled words and abbreviations. In this way, FastText converts the words often ignored by classical preprocessing pipelines into known entities. Such features help in processing free-form text as it contains abbreviations and short forms of words. These non-contextualized embedding-based vectors preserve the semantic meaning with a certain degree of underlying language understanding. The main disadvantage of non-contextualized word embedding is that they are static, meaning each word has a fixed vector representation regardless of the context. For instance, resulting vectors can capture relations between words, such as vector for "king" – "man" + "woman" results into vector close to "queen" (gender relationship). The vector for "Paris" – "France" + "Poland" results in a vector close to "Warsaw" (capital relationship). But for polysemic words such as "bank", when used as river bank and financial bank, will have the same fixed embedding regardless of the context it is used in. This drawback led the research towards the development of techniques to generate contextualized word embedding using techniques like recurrent neural networks [10,14] and attention mechanisms [4,20]. These efforts led to the development of famous NLP models like ELMo [15], OpenAI GPT [16] and BERT [4]. BERT introduced a novel training methodology based on transformers architecture that takes advantage of the attention mechanism presented in [20]. This bidirectional learning approach achieved very good performance, pushing state-of-the-art on 11 downstream tasks including classification, language inference, Named Entity Recognition, question-answering, and others. After the release of BERT as an open-source technique, different NLP groups were able to produce similar state-of-the-art results on different languages and tasks [21]. One of the main advantages behind the success of BERT is transfer learning - the ability to adopt large pre-trained deep learning models across different languages, tasks and domains [17]. This feature of BERT accelerated progress in the field of NLP that ImageNet brought to the computer vision community.

This study focuses on predicting problem categories in maintenance records written in French language. We used the French version of BERT (CamemBERT) which is based on RoBERTa's architecture and achieved state-of-the-art results on various French NLP tasks. This model is trained on 138 GB of uncompressed French text through 17 h of GPU training [9]. Literature analysis shows that the use of contextualized word embedding is not very common on free-form text especially under the context of Industry 4.0. To the best of the author's knowledge, this is the second time CamemBERT is used on free-form industrial dataset, with the first study being [19]. The remainder of this paper is organized as follows: Sect. 2 will discuss the overall methodology : data preprocessing and augmentation steps, and ML architectures used to develop the classification system. Section 3 will describe the results in detail and Sect. 5 will present the directions for future work. Finally, paper is concluded in the Sect. 4.

2 Methodology

This section presents the dataset, along with a description of preprocessing steps. Also, the architecture and methodology used to develop the classification system are explained.

2.1 Dataset Preprocessing

Dataset used in this study is obtained from a manufacturing company. Due to confidentiality constraints, the name and industry are not mentioned in this paper. The dataset contains maintenance records collected over the years from 2015 to 2021. These records contain a total of 814 entries classified into four problem categories: *mechanical*, *electronic*, *electrical* and *miscellaneous*. The miscellaneous category consists of samples from three different categories (information technology, network, and pneumatic) combined into a single category due to lower sample counts. The details of the available variables are described in Table 1 and Fig. 1 shows a sample of maintenance records before and after preprocessing.

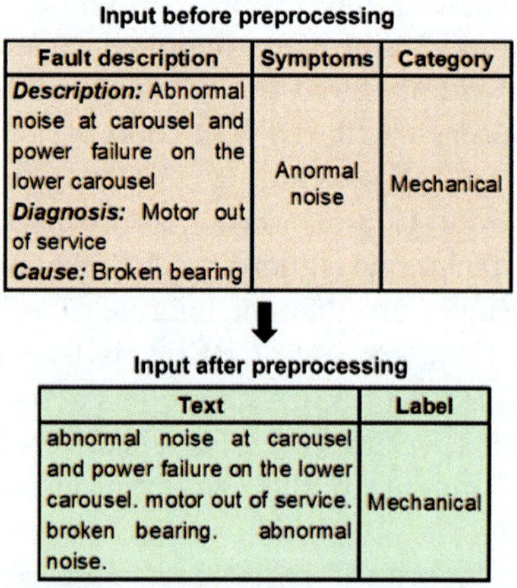

Fig. 1. Sample maintenance record before and after preprocessing

Table 1. Description of dataset used for study

Attribute	Data type	Details
Fault description	String (*free-form text*)	Description of fault, diagnostic steps involved and cause of the fault
Symptoms	String (*free-form text*)	Description of symptoms of the fault
Category	String (*categorical*)	Four problem categories: 1. Mechanical 2. Electronic 3. Electrical 4. Miscellaneous

One of the main reasons behind the selection of deep contextualized embedding is the minimal preprocessing requirement. This saves a lot of time designing preprocessing pipelines using different NLP toolkits. BERT can process the text in its raw form. In our case, the fault description variable was in two different formats: it either contained the problem description without tags or problem description with tags (description further divided into the description, diagnostic, and cause). As most of the problem descriptions are without tags it was decided to normalize all the descriptions into the same format: as description without tags. To do that, tags were removed to make a single continuous problem description. Another preprocessing step performed is the concatenation of symptoms at the end of each description to generate a single input sequence for each maintenance record. The concatenation of symptoms with description was decided due to the high percentage of missing values in symptoms.

2.2 Dataset Distribution, Augmentation and Splitting

In Table 2 it can be observed that there is an imbalance between the sample count of different classes. To overcome this issue data augmentation is applied.

Table 2. Distribution of samples among different problem categories

	Mechanical	Electronic	Electronic	Miscellaneous	**Total**
No. of samples	308	234	143	129	**814**

Data augmentation has been widely used in computer vision but it is also gaining popularity in the NLP community for studies where resource data is low or class imbalance poses difficulties in training and evaluation phases.

Another reason behind the popularity of data augmentation in the NLP community is the introduction of very large-scale transformer-based neural network models that require large amounts of data for training [5]. After preprocessing step, to avoid the introduction of bias in the model, data is split into train and test sets before augmentation. Figure 2 shows the flow diagram of data splitting and augmentation. A train-test split of 90%-10% is used in order to maximize the amount of training data. This is done because deep contextualized embedding models, like BERT, need large quantities of data.

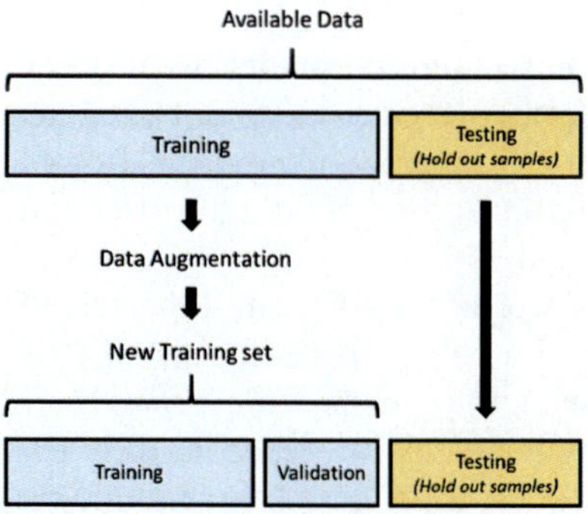

Fig. 2. Flow diagram of data splitting and augmentation

Table 3. Distribution of training and test samples with and without augmentation

Category	Training samples		Test samples
	Without augmentation	With augmentation	Holdout samples
Mechanical	277	277	31
Electronic	208	277	26
Electrical	133	277	10
Miscellaneous	114	277	15
Total	**732**	**1108**	**82**

To obtain a balanced training set and increase training data, data augmentation was applied using the *nlpaug* Python package [8]. This package provides different augmentation strategies including character, word, and sentence level augmentation. For each of these strategies, augmentation is performed through random substitution, random deletion, random insertion, shuffling, and synonym replacement, etc. *nlpaug* package also supports augmentation using classical techniques like TF-IDF, Word2vec and also has the capability of augmentation using more advanced NLP techniques such as BERT and XLnet based contextualized word embedding. As our study is focused on the role of contextualized embedding based on BERT, we also used this technique for augmentation. Using CamemBERT-large masked language model, a random substitution was performed using contextualized word embedding. As augmentation introduces some unwanted noise and there is a chance that the new data is very different from the actual data it was decided to keep the augmentation to a minimum. Data augmentation was therefore only applied to less frequent classes (electronic, electrical, and miscellaneous) compared to the most frequent class (mechanical).

To augment each class with an equal number of samples, training data is divided into separate classes (mechanical, electronic, electrical, and miscellaneous). Then for every class (except mechanical), each data sample is used to generate five augmented samples. Finally, the required number of randomly selected augmented samples were added to original training examples to generate a new training set with an equal number of samples (277 samples) per class. Table 3 shows the distribution of samples among training and test with and without augmentation. To avoid bias in the model, the test set is kept the same as the one obtained through data splitting before augmentation. After creating the new training set, different ML models were trained using various feature combinations. In the next section different ML models trained to develop the classification system are discussed in detail.

2.3 Machine Learning Models

To test the performance of various classification techniques, models were trained using five different approaches. To see the performance gain due to data augmentation, separate models were trained with and without augmentation. In addition, to show the effectiveness of deep contextualized embedding on free-form text, deep contextualized embedding-based models are compared with models trained using classical features such as bag of words (BOW) and TF-IDF. TF-IDF is one of the most used methodologies for text vectorization in NLP [1]. Figure 3 shows an overview of model development pipeline from input to the selection of best model, including different types of features and models used in the study. Different approaches used are discussed in detail in the following subsections.

Baseline Model: To test the performance of all models with a baseline, a dummy model was developed. This model only serves as the baseline against other models and can not be used as the final model for classification. The approach used to develop the dummy model was predicting the most frequent class that is *mechanical* in our case.

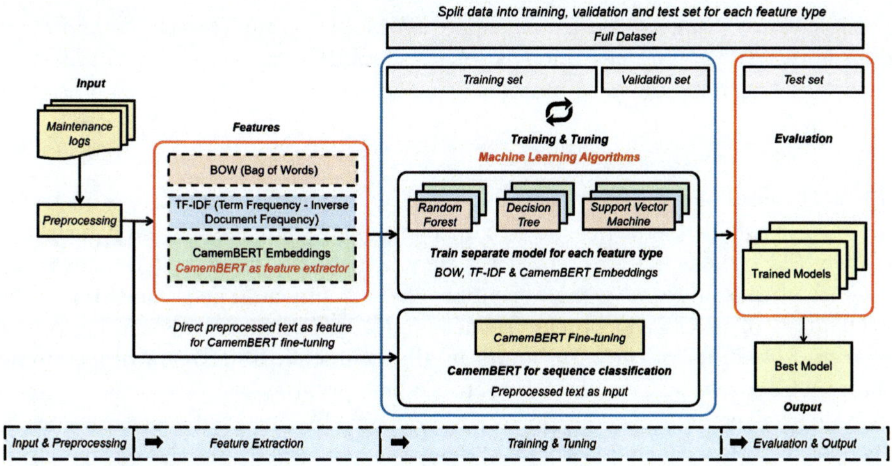

Fig. 3. Model development pipeline from input to output (best model) - *same pipeline is followed for training data with and without augmentation.*

Hyperparameter Tuning: The performance of a machine learning model significantly depends on the chosen hyperparameters. Manual hyperparameter tuning is time-consuming and it is difficult to try so many combinations. To avoid manual tuning and automate this process, grid search for hyperparameters was applied. To use grid search we used *GridSearchCV* function of *sklearn's model_selection package.* Using this function, we can identify the best set of hyperparameters to fit the model on the training set. To find the best set of hyperparameters we provide *GridSearchCV* function with a dictionary of hyperparameters containing hyperparameters and possible values the hyperparameters can take. Then *GridSearchCV* function tries all the possible combinations of these parameters and evaluates the model for each combination using cross-validation. In the end, the best parameters can be selected for the final model by comparing the performance metrics. Only the hyperparameters in the provided dictionary are tuned and the default value of the rest of the hyperparameters are used. Table 4 shows the hyperparameters along with provided value ranges for the different models used in the grid search. Hyperparameter tuning of CamemBERT fine-tuning-based models is discussed in Sect. 2.3.

Bag of Words Feature Based Models: To convert training data into classifier input, each sample needs to be converted into a numeric vector. In the BOW approach, a vocabulary is generated that is a list of all the unique words in the data set. Using this vocabulary, a fixed numerical vector is created for each training sample, using the count of each unique word in the given training example. To reduce the size of vocabulary and vectors, stop words were removed from the data before vectorization. After generating features for both training sets (with and without augmentation), three different types of models are trained using these features: Support Vector Machine (SVM), Decision Tree (DT), and Random Forest (RF). The best models for these models types are chosen based on the best performing hyperparameters using the grid search parameters presented in Table 4.

TF-IDF Feature Based Models: TF-IDF is the product of two measures i.e. term frequency, and inverse document frequency. In the first step term frequency (TF) is calculated. Term frequency, for a given document measures the number of times a term (word), occurs in that document. The second step is the calculation of inverse document frequency (IDF). IDF measures the log of the inverse probability of the term found in all documents [18]. To convert training samples from raw text to TF-IDF vectors, some preprocessing steps are required including tokenization and stop words removal. In this study, performance is evaluated both with and without stop words removal to see the effect of this preprocessing step on the performance of the models, especially in the context of maintenance text. The maximum feature size is fixed to 1024 which is the same size as deep contextualized embedding generated by CamemBERT-large model. Similar to the BOW strategy, the performance of TF-IDF features is also evaluated on training sets with and without augmentation. Furthermore, the same classification methods SVM, DT, and RF are trained. Best models are identified by choosing the best performing models through grid search. The search space for these parameters is presented in Table 4.

Table 4. Range of hyperparameters used for different models to apply grid search

Model	Hyperparameters for grid search
SVM *(Support vector machine)*	C : [1, 10, 100, 1000]
	$Kernel$: $[linear, rbf]$
	$gamma$: [1e − 3, 1e − 4]
DT *(Decision Tree)*	$criterion$: $[gini, entropy]$
	max_depth : [4, 5, 6, 7, 8, 9, 10, 11, 12, 15, 20, 30, 40, 50, 70, 90, 120, 150]
RF *(Random Forest)*	$bootstrap$: $[True, False]$
	$max\_features$: $[auto, sqrt]$
	$min\_samples\_leaf$: [1, 2, 4]
	$min\_samples\_split$: [2, 5, 10]
	max_depth : [10, 20, 30, 40, 50, 60, 70, 80, 90, 100, 110, $None$]
	$n\_estimators$: [200, 400, 600, 800, 1000, 1200, 1400, 1600, 1800, 2000]

CamemBERT Fine-Tuning Based Models: Despite low time complexity and simplicity, both BOW and TF-IDF have two major disadvantages. Firstly, with an increase in dataset size, the number of unique words increases, resulting in much larger vectors. Both these methods being frequency-based techniques, vector size is dependent on the number of unique words in the vocabulary. Secondly, these techniques do not handle the order in which the words appear and only focus on the word frequency. That is why it is important to explore recent transformer-based state-of-the-art techniques in NLP like BERT. BERT provides fixed-size contextualized embedding that can be adopted for various downstream tasks, like sequence classification in our case. To adopt the BERT model trained on general language to a specific domain, fine-tuning needs to be performed. To perform domain-specific fine-tuning of BERT-based models, a feed-forward neural network with softmax is jointly trained using pooled BERT output from the pre-trained BERT model. This helps adapt the contextualized embedding previously trained on generalized language data to the target domain using minimal data, time, and computational resources. This approach also requires minimal architecture modification to achieve state-of-the-art performance on the given task. Figure 4 shows the architecture of domain-specific fine-tuning of BERT-based models (CamemBERT in our case).

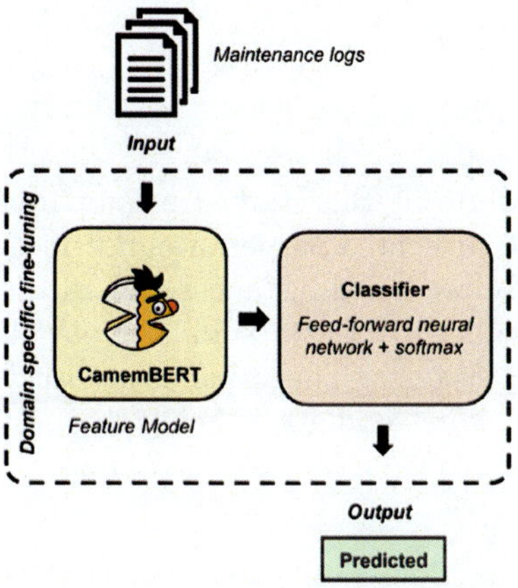

Fig. 4. Architecture of domain specific fine-tuning for the BERT based model

Table 5. Best performance model parameters for CamemBERT fine-tuned models

CamemBERT fine-tuned	Best Hyperparameters
Training *(without augmentation)*	$Epochs$: 4 $Batchsize$: 16 $Warmupsteps$: 0 $Optimizer$: $Adam$ ($LearningRate$: 5×10^{-5}) $Maxsequencelength$: 128
Training *(with augmentation)*	$Epochs$: 3 $Batchsize$: 16 $Warmupsteps$: 0 $Optimizer$: $Adam$ ($LearningRate$: 2×10^{-5}) $Maxsequencelength$: 128

To train a domain-specific fine-tuned model for free-form maintenance text, CamembertForSequenceClassification class available in *Transformers* Python library was used [21]. This class takes a pre-trained BERT model and the total number of classes as input and adds a feed-forward neural network with a softmax on top of it. Two different versions of this model were trained during system development, one for each training set (with and without augment). Different hyperparameter ranges tested for optimization include: epoch (1, 2, 3 and 4), batch size (8, 16 and 32), max sequence length (64, 96 and 128) and learning rate (2×10^{-5}, 3×10^{-5}, 4×10^{-5} and 5×10^{-5}. The best performance was achieved using hyperparameters presented in Table 5.

CamemBERT Feature Based Models: In addition to fine-tuning the CamemBERT model on domain-specific data by adding a neural network with a softmax layer on top of it, we can also use the CamemBERT pre-trained model to generate contextualized word embedding. In BERT-based models, the process of converting raw text to the input sequence adds some extra special tokens to the sequence by the inbuilt model tokenizer. One of these special tokens is [CLS]. The embedding of [CLS] token serves as an aggregated representation of the input sequence for classification tasks. These embeddings can be used as features to train some other ML model on top of pre-trained or fine-tuned BERT model [4]. This analysis is performed to compare the effect of contextualized embedding-based features on the same models trained using BOW and TF-IDF-based features. The features obtained from these fine-tuned models consist of fixed size embedding of size 1024. Using embedding from CamemBERT models, other models were trained including SVM, DT, and RF. The hyperparameters of these models are tuned using gird search for provided parameter ranges in Table 4.

3 Results and Discussion

In this section results of various models trained during this study are compared on the test set. Accuracy and MCC score are presented as performance metrics. Accuracy is the ratio between the correctly predicted instances and all the instances in the test set. Many researchers use accuracy as a standard way to estimate the performance of machine learning systems. The value of accuracy ranges between 0 and 100%, with 0 being the worst and 100% being the best accuracy. Accuracy can be calculated using Eq. 1.

$$Accuracy = \frac{TP + TN}{TP + TN + FP + FN} \tag{1}$$

Where TP (true positive) is the number of correctly labelled positive samples, TN (true negative) is the number of correctly labelled negative samples, FP (False Positive) is the number of negative samples incorrectly labelled as positive and FN (false negative) is the number of positive samples incorrectly labelled as negative. MCC score on the other hand is used to provide insight into how well the class imbalance problem was handled. This is because MCC produces a high score only if the prediction obtained good results in all four confusion matrix categories, such as TP, FN, TN, and FP. MCC score ranges between -1 and 1, with -1 being the worst and 1 being the best score. MCC score can be calculated using Eq. 2.

$$MCC = \frac{(TP \times TN) - (FP \times FN)}{\sqrt{(TN + FP) \times (TP + FN) \times (TN + FP) \times (TN + FN)}} \tag{2}$$

3.1 Performance with and Without Data Augmentation

Various models are trained during the study. Table 6 presents test accuracy and MCC score of various models trained using non augmented and augmented data. Results of the best model are highlighted in the table.

Table 6. Results of various models on test set (training with and without data augmentation)

Features	**Model**	**Non augmented**		**Augmented**	
		Accuracy	**MCC**[f]	**Accuracy**	**MCC**
BoW[a]	SVM[c]	59.76	0.43	60.98	0.46
	DT[d]	58.54	0.42	52.44	0.36
	RF[e]	71.95	0.61	64.63	0.50
TF-IDF[b] *(with stopwords)*	SVM	70.73	0.58	65.85	0.53
	DT	52.44	0.32	51.22	0.32
	RF	69.51	0.57	70.73	0.59
TF-IDF *(without stopwords)*	SVM	65.85	0.51	64.63	0.50
	DT	57.32	0.40	60.98	0.45
	RF	68.29	0.55	67.07	0.54
CamemBERT Embedding	SVM	73.17	0.63	78.05	0.70
	DT	70.73	0.59	69.51	0.57
	RF	74.39	0.65	76.83	0.68
Preprocessed Text	CamemBERT-Large *with* FFNN[g] + Softmax ↓ *CamemBERT fine-tuned*	**80.21**	**0.69**	**82.29**	**0.71**

[a]*BoW: Bag of Words,* [b]*TF-IDF: Term Frequency - Inverse Document Frequency*
[c]*SVM: Support Vector Machine,* [d]*DT: Decision Tree,* [e]*RF: Random Forest*
[f]*MCC: Matthews Correlation Coefficient,* [g]*FFNN: Feed-forward Neural Network*

Results indicate that the CamemBERT model with fine-tuning approach has the best performance with respect to both accuracy and MCC score. It can also be observed that all the models are performing better using contextualized embedding compared to other features, both in terms of accuracy and MCC score.

Similarly, Table 6 also presents the accuracy and MCC score of various models after data augmentation. Results indicate an increase in the accuracy and MCC score for the CamemBERT model with fine-tuning approach after data augmentation. However, in both augmented and non augmented data cases CamemBERT with fine-tuning approach is the best performing model.

In summary results show that deep contextualized embedding can be used to develop ML applications using free-form text. Although results of the CamemBERT model with fine-tuning approach (best model) were acceptable without data augmentation, results indicate that data augmentation helps to reduce the effect of class imbalance. Furthermore, deep contextualized embedding when used as a feature in classical machine learning algorithms (SVM, RF, and DT) also produce better results compared to classical feature techniques such as BOW and TF-IDF.

4 Conclusion

This study explored the capability of the state-of-the-art BERT-based French NLP model, CamemBERT on free-form maintenance text. The purpose of the study was to estimate the category of maintenance problems using log entries by maintenance operators. CamemBERT was used to categorize the text using domain-specific transfer learning both in fine-tuning and feature-based mode. The dataset used for the study is a real industrial dataset. The provided data set contained highly unstructured text with class imbalance; as some problem categories are more common compared to others. To handle the class imbalance problem data augmentation was performed using word-level random substitution with CamemBERT based contextualized word embedding. Findings indicate that CamemBERT with a fine-tuning approach resulted in the best accuracy and MCC score on the test set using both training sets (with and without augmentation). Similarly, it can be observed that other models (SVM, DT, RF) when trained using contextualized embedding-based features gave better performance compared to classical features like BOW and TF-IDF. The best model was able to achieve test accuracy of 80.21% with an MCC of 0.69 when trained using the training set without augmentation and accuracy of 82.29% and MCC of 0.71 when trained using the training set with augmentation. This indicates data augmentation helps increase model performance by handling class imbalance. The study also shows using BERT-based state-of-the-art models we can effectively explore the text in maintenance logs to develop applications for industrial processes improvement.

5 Future Work

In the context of the Technical language Processing (TLP) paradigm and industry 4.0 [3], there is a need for similar studies to adopt NLP pipelines for free-form maintenance text. In continuation with this study future work will focus on the following directions:

- This study is the initial work towards the development of maintenance applications based on free-form maintenance text to facilitate engineers to efficiently perform maintenance tasks.
- Future work will also include the comparison between the performance of similar French BERT-based models like FlauBERT to analyze which model performs the best on free-form maintenance text.
- Performance analysis on other similar industrial datasets to identify generalized approaches to handle this type of data. Also, various data augmentation techniques need to be explored and tested on multiple datasets to identify the best approach for free-form maintenance text.

Acknowledgments. This work has been supported by the EIPHI Graduate school (contract "ANR-17-EURE-0002").

References

1. Beel, J., Gipp, B., Langer, S., Breitinger, C.: Paper recommender systems: a literature survey. Int. J. Digital Libr. **17**(4), 305–338 (2016)
2. Bojanowski, P., Grave, E., Joulin, A., Mikolov, T.: Enriching word vectors with subword information. Trans. Assoc. Comput. Linguist. **5**, 135–146 (2017)
3. Brundage, M.P., Sexton, T., Hodkiewicz, M., Dima, A., Lukens, S.: Technical language processing: unlocking maintenance knowledge. Manuf. Lett. **27**, 42–46 (2021)
4. Devlin, J., Chang, M.W., Lee, K., Toutanova, K.: BERT: pre-training of deep bidirectional transformers for language understanding. arXiv preprint: arXiv:1810.04805 (2018)
5. Feng, S.Y., et al.: A survey of data augmentation approaches for NLP. arXiv preprint: arXiv:2105.03075 (2021)
6. Hirsch, D.D.: The glass house effect: Big data, the new oil, and the power of analogy. Me. L. Rev. **66**, 373 (2013)
7. Jones, K.S.: A statistical interpretation of term specificity and its application in retrieval. J. Doc. **60**, 493–502 (1972)
8. Ma, E.: NLP augmentation (2019). https://github.com/makcedward/nlpaug
9. Martin, L., et al.: CamemBERT: a tasty French language model. arXiv preprint arXiv:1911.03894 (2019)
10. McCann, B., Bradbury, J., Xiong, C., Socher, R.: Learned in translation: contextualized word vectors. arXiv preprint: arXiv:1708.00107 (2017)
11. Mehta, Y., Majumder, N., Gelbukh, A., Cambria, E.: Recent trends in deep learning based personality detection. Artif. Intell. Rev. **53**(4), 2313–2339 (2019). https://doi.org/10.1007/s10462-019-09770-z
12. Mikolov, T., Sutskever, I., Chen, K., Corrado, G., Dean, J.: Distributed representations of words and phrases and their compositionality. arXiv preprint: arXiv:1310.4546 (2013)
13. Pennington, J., Socher, R., Manning, C.D.: Glove: Global vectors for word representation. In: Proceedings of the 2014 Conference on Empirical Methods in Natural Language Processing (EMNLP), pp. 1532–1543 (2014)
14. Peters, M.E., Ammar, W., Bhagavatula, C., Power, R.: Semi-supervised sequence tagging with bidirectional language models. arXiv preprint: arXiv:1705.00108 (2017)
15. Peters, M.E., et al.: Deep contextualized word representations. arXiv preprint: arXiv:1802.05365 (2018)
16. Radford, A., Narasimhan, K., Salimans, T., Sutskever, I.: Improving language understanding by generative pre-training (2018)
17. Ruder, S.: Neural transfer learning for natural language processing. Ph.D. thesis, NUI Galway (2019)
18. Tian, X., Tong, W.: An improvement to TF: term distribution based term weight algorithm. In: 2010 Second International Conference on Networks Security, Wireless Communications and Trusted Computing, vol. 1, pp. 252–255. IEEE (2010)
19. Usuga Cadavid, J.P., Grabot, B., Lamouri, S., Pellerin, R., Fortin, A.: Valuing free-form text data from maintenance logs through transfer learning with camembert. Enterprise Information Systems, pp. 1–29 (2020)
20. Vaswani, A., et al.: Attention is all you need. arXiv preprint: arXiv:1706.03762 (2017)
21. Wolf, T., et al.: Huggingface's transformers: State-of-the-art natural language processing. arXiv preprint: arXiv:1910.03771 (2019)

Context-Aware Explanations in Recommender Systems

Jinfeng Zhong(✉) and Elsa Negre

Paris-Dauphine University, PSL Research University, CNRS UMR 7243, LAMSADE 75016, Paris, France
jinfeng.zhong@dauphine.eu, elsa.negre@lamsade.dauphine.fr

Abstract. Recommender systems aim to help users find relevant items more quickly by providing personalized recommendations. Explanations in recommender systems help users understand why such recommendations have been generated, which in turn makes the system more transparent and promotes users' trust and satisfaction. In recent years, explaining recommendations has drawn increasing attention from both academia and from industry. In this paper, we present a user study to investigate context-aware explanations in recommender systems. In particular, we build a web-based questionnaire that is able to interact with users: generating and explaining recommendations. With this questionnaire, we investigate the effects of context-aware explanations in terms of efficiency, effectiveness, persuasiveness, satisfaction, trust and transparency through a user study.

Keywords: Context-aware explanations · Explainable recommendations · Recommender systems

1 Introduction

With the increasingly available information online, users are sometimes overwhelmed by information overload. Recommender systems (RSs hereinafter) have become important decision-support tools that aim to help users find relevant items (goods, services, etc.). Among the existing recommender techniques, Context-Aware Recommender Systems (CARSs hereinafter) that leverage users' contextual information [1] are able to provide more personalized recommendations. According to [2], context is any information that can be used to characterize the situation of an entity. In RSs, users' contextual situations are composed of corresponding contextual conditions. For example, users' contextual situations can be composed of location, time and companions, etc. Indeed, users' contextual information (e.g. location, moods, time) can greatly influence their preferences. For example, in a movie recommendation scenario, a romantic movie is more likely to be preferred when users are with their lovers and a cartoon movie seems to be a better recommendation when users are with their children. In this

L. Troiano et al. (Eds.): ICDLAIR 2021, LNNS 441, pp. 76–85, 2022.
https://doi.org/10.1007/978-3-030-98531-8_8

toy example, users' contextual information (companion) is leveraged to better model their preferences. CARSs have achieved great success for being able to leverage such contextual information [3]. However, as recommender algorithms become more and more inscrutable, many RSs work like black boxes, raising concerns about transparency and fairness. Explaining recommendations to help users better understand how RSs work has drawn much attention from both academia and industry.

Since users' preferences vary across different contextual situations, it would be logical to provide explanations to users based on their contextual situation, which may make the explanations more personalized and more convincing. Therefore, context-aware explanations such as "It would be nice to watch this movie with your family" would be much more personalized and convincing than traditional explanations such as "We recommend this film because the average rating of this movie is high".

Tintarev et al. [4] asserted that explanations can make RSs more transparent (transparency), help users make better decisions (effectiveness) more quickly (efficiency), convince users to consume (persuasiveness), increase users' confidence in the system (trust) and increase the ease of use or enjoyment (satisfaction). As a matter of fact, the potential goals can also serve as metrics for evaluating explanations in RSs [5]. In this paper, we develop a web-based questionnaire that can generate and explain recommendations. With this questionnaire, we carry out a user study to compare the effects of explanations with and without contextual information in terms of transparency, effectiveness, efficiency, persuasiveness, trust and satisfaction.

The remainder of the paper is structured as follows. In Sect. 2, we summarize related works about CARSs, explanations in recommendations, the metrics of evaluating explanations in RSs. In Sect. 3, we present in detail the problem formulation and the user study design. In Sect. 4, we present the data collected and corresponding analysis. Lastly, we conclude and propose potential future works.

2 Related Works

2.1 Context-Aware Recommender Systems (CARSs)

CARSs can generate more accurate and personalized recommendations by leveraging users' contextual situations [1]. CARSs assume that users' preferences change over different contextual situations. For example, when recommending movies, users' preferences can be influenced by companions (with family or alone), mood (happy or sad). The "companion" and "mood" are contextual factors, each of which can have several possible values (contextual conditions). Finally the contextual conditions compose contextual situations. Therefore, CARSs differ from traditional RSs (e.g. collaborative filtering) in the dimension of data on which they operate. Typical data records in collaborative filtering are $< user, item, rating >$ while in CARSs data records are in the form of $< user, item, contexts, rating >$. The high dimension of data also makes models in CARSs more complex. The rating prediction function in CARSs is multidimensional: $user \times item \times contexts \rightarrow rating$. In a word, the high dimension of data and complexicity of models lead to the fact that CARSs have become more inscrutable. Considering that users' contextual situations influence their interests, explaining context-aware recommendations is worth extensive exploration [6]. In the next section, we review existing works related to context-aware explanations in RSs.

2.2 Context-Aware Explanations in Recommender Systems (RSs)

Depending on the information included in explanations, existing explanation methods in RSs can be classified into similar user (item) explanation, feature-based explanation, opinion-based explanation and social explanation [6]. However, these explanations do not consider users' contextual situations, which means that explanations under different contextual situations are the same. Therefore, traditional explanations may not be adapted to users' needs under different situations. Contextual information, which can influence users' preferences, has not been sufficiently explored to generate explanations [7]. Baltrunas et al. [8] showed that, compared with non-contextual explanations, context-aware explanations can promote user satisfaction when recommending places of interest. One major concern is that only 20 participants were involved in their user study. Misztal et al. [9] conceptualized a CARS that can explain recommendations by identifying the most relevant context for each movie type. However, explanation for items of the same type are not diversified and the proposed conceptualization was not put into application. Sato et al. [10] showed through a user study that context-aware explanation can promote persuasiveness and usefulness, other metrics such as efficiency and trust were not evaluated. More recently, Musto et al. [7] proposed context-aware natural language explanations. Recommendations were generated by a content-based approach according to descriptive features of items, explanations are generated by exploiting distributional semantics in reviews posted by other users. Their user study revealed that context-aware explanations were preferred to non-contextual explanations. However they did not further investigate other effects of context-aware explanations.

We note that there are a few works related to contextual explanations in RSs as presented above. In this work, we comprehensively explore the effects of context-aware explanations in RSs. The traits that distinguish our work from the works mentioned above are: (1) Participants can provide ratings given several contextual situations, allowing our system to better learn their preferences, which is not the case in [7–10]; (2) Contexts in [7] are not used in recommender models, therefore the context-aware explanations may not reflect the real reason why users like a recommendation. On the contrary, explanations in our system are explicit, the most influential contextual conditions serve as explanation; (3) Our questionnaire is able to investigate the effects of context-aware explanations by fully evaluating the six metrics (efficiency, effectiveness, persuasiveness, satisfaction, trust and transparency) presented in Sect. 1, allowing us to evaluate explanations from a broader point of view.

3 Our Proposition

3.1 Recommendation Method

As we presented Sect. 2, context-aware explanation worth extensive exploration. In this work, we aim to fully explore the effects of context-aware explanations in recommender systems (RSs) in terms of efficiency, effectiveness, persuasiveness, satisfaction and transparency.

We design a web-based questionnaire that is able to interact with participants: recommending and explaining recommendations. To generate context-aware recommendations, we applied a method we recently proposed in [11]. The main steps mainly include: (1) Cluster items (movies) using Hierarchical Clustering as [12]; (2) Calculate the item-based Pearson Correlation Coefficient (PCC) [13] between each contextual condition and ratings for each item cluster; (3) Represent each contextual condition based on the calculated PCC; (4) Calculate the similarity between the target contextual situation CS^* and the contextual situations CS existing in the original dataset; (5) Select a local dataset whose contextual situations are similar to the target contextual situation according to a similarity threshold; (6) A traditional 2D recommender technique is applied in the local dataset to obtain a set of recommendations. For more details, we refer to [11].

3.2 Experiment Setup

We adopt the widely used contextual movie recommendation dataset CoMoDa [14], where there are 121 users, 1197 movies, 2296 ratings and 12 contextual factors. In the CoMoDa dataset, the average number of ratings of each user is about 12. In order to be consistent with the original dataset, participants are required to give their ratings for 12 movies under different contextual situations. In the original dataset, there are 12 contextual factors, some contextual factors could have a more important impact than others [15]. We wish to select the most impacting contextual factors to minimize the efforts of participants to specify his/her contextual situation. According to [15], in the CoMoDa dataset, the 6 most impacting contextual factors are *(i) whether users have watched a movie before, (ii) users are given a decision or decide themselves, (iii) users' physical wellness, (iv) users' moods, (v) users' location and (vi) weather.* Note that in step 2 of our protocol, we recommend movies that participants have not interacted with in step 1 of our protocol, we do not consider the contextual factor *Interaction*. Since all the participants are given a movie, we do not consider the contextual factor *Decision* either. Therefore, in our case, we only consider the following 4 contextual factors: *Physical wellness*, *Mood*, *Location* and *Weather*.

There are mainly four steps in the user study: (1) Participants are asked to rate 12 randomly selected movies given several contextual situations by applying the data set in [14]. This step is to learn users' preferences under different contextual situations; (2) The system generates recommendations for the participants under a randomly generated contextual situation, for more details we refer to [11]; (3) Participants are presented with explanations but without

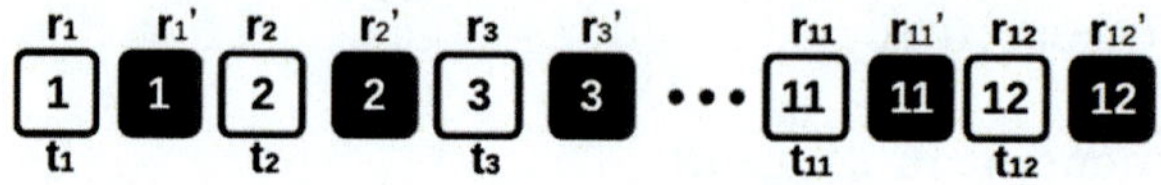

Fig. 1. The number indicates the movie recommended. White indicates that only explanation is displayed, black indicates that detailed information of the movie is displayed, r is the rating given by participants, and t is the time cost by participants to give a rating.

detailed information of movies and they are asked to rate the movie, noted as r, the time consumed for giving the rating is noted as t, users are then presented with detailed information of the same movie and rate them again, noted as $r^{'}$, Fig. 1 illustrates this process; (4) Participants are asked to evaluate the explanations in terms of transparency, effectiveness, efficiency, persuasiveness, trust and satisfaction by a 1 (lowest) to 5 (highest) rating scale. Therefore, the six metrics can be quantified as below: (1) Efficiency: the time used by participants for giving a rating when only presented with explanations; (2) Effectiveness: the difference between r and $r^{'}$, if the difference is near 0, it indicates good effectiveness since explanation has helped participants perfectly estimate the rating of recommendation; (3) Persuasiveness: $r - r^{'} > 0$ means positive persuasiveness; $r - r^{'} < 0$ means negative persuasiveness; (4) Transparency, satisfaction and trust: they are reflected by users' responses in step 4; (5) In terms of efficiency, effectiveness and persuasiveness, we will compare the measured effects and human perceived effects. The protocol adopted in this user study basically originated from [16]. The protocol was adopted by [5] where the authors studied how explanation style can influence effects of different explanation styles in terms of efficiency, effectiveness, persuasiveness, transparency, satisfaction. Our protocol differs from that of [5] in that: (1) We measure the effects of explanations in terms of trust, which is not considered in [5]; (2) In terms of efficiency, effectiveness and persuasiveness, we also ask the user again to evaluate different styles of explanations in terms of these three metrics to verify if there are differences between the human perceived feelings and actual effects while in [5], these three metrics are only measured by objective methods.

3.3 Baselines

Concretely, in this paper, we wish to compare the effects of explanations with and without contextual information. Table 1 presents the explanation types we plan to show to participants in the experiment. The *Avg* style explanation shows users the average rating of the recommended item; *Per* style explanation presents the distribution of ratings of the recommended item; *Simu* style explanation presents the average ratings of similar users. *Simu*, *Per* and *Avg* style explanations are among the 21 explanation styles that Herlocker et al. [17] compared in terms of persuasiveness. Since explanations that perform well in one goal may perform poorly in another (trade-off among goals) [4] and we investigate six different goals, we do not simply select the top ranking explanation types in [17]. We selected the top, middle and end of the ranking list of the 21 explanation styles, namely *Simu*, *Per*

Table 1. Explanations to be presented.

Name	Example
Avg	The average rating of this movie is 3.8
Per	80% of users rate this movie more than 4
Simu	The average rating of users whose preferences are similar with yours is 4.1
Simi	This movie is similar to movies you watched before
Content	This is a movie directed (acted) by
Context-aware	**The system suppose that you would like to watch this movie when it shines, healthy you are at home and in good moods**

and *Avg* respectively. *Simi* style explanation simply says the recommended item are similar to the items a participant has interacted before, which is popular in web sites such as Amazon [18]; *Content* style explanation presents the characteristics of recommended items that may be preferred by users, it is widely adopted for movie, book recommendation [19]. *Context-aware* explanation leverages users' contextual information to generate explanations [10]. In order that these explanation styles can be compared fairly, they are all in textual forms.

4 Results of Experiments

4.1 Statistics About the Participants and Data Obtained

Since the questionnaire is web-based, we were able to send it to researchers in and out of our laboratory, students in the university through mail lists. The questionnaire was online for one week and the participants can answer it anonymously. Overall, we have received 147 responses. In terms of sex, female and male each occupied nearly half. Most of the participants age from 18 - 55; over 66% of them have an education level of more than Bac+5 (master level); nearly half of the participants are student (ranging from bachelor to PHD students), employee of companies and higher education occupy nearly another half, this is mainly due to the fact that most of the people in the mail lists are researchers or students; more than half of the participants watched movies regularly, indicating high interest in movies.

4.2 Observations of Users' Responses

In this section, we present the analysis obtained from users' responses. We note that in order to analyze the observed differences between different explanation types, we first applied analysis of variance [20]. If significant differences were found, we then applied Tukey HSD test [21] to identify the specific differences. All the tests were done at a 95% confidence level.

Table 2. Efficiency, effectiveness and persuasiveness of different types of explanations. Efficiency is reflected by the time consumed by participants to give a rating (in seconds); effectiveness is reflected by the difference between r and $r^{'}$ (as presented in Fig. 1), the more close it is to 0 the more effective an explanation is; persuasiveness is also measured by the difference between r and $r^{'}$, $r - r^{'} > 0$ means positive persuasiveness; $r - r^{'} < 0$ means negative persuasiveness.

	Efficiency (seconds)	Effectiveness and persuasiveness
Avg	4.78	0.30
Content	5.37	**−0.02**
Context-aware	**8.25**	**−0.01**
Per	5.17	0.11
Simi	5.99	−0.18
Simu	6.89	0.10

Objective Evaluation. In this section, we will present the results of objective evaluation: efficiency, effectiveness and persuasiveness. Table 2 shows the results of these three metrics.

It can be observed that, when presented with *Context-aware* explanations, participants spend more time making a decision (See row *Context-aware* and column *Efficiency (seconds)*). In our case, making a decision means giving a score. Although there are not significant differences among different explanation types in terms of $r - r^{'}$, *Content* style and *Context-aware* explanations help participants slightly better estimate the quality of recommendations because the difference between r and $r^{'}$ is close to zero. *Avg* and *Simi* explanation types make users underestimate the qualities of recommendations. *Per* and *Simu* style explanations are slightly more persuasive than other styles. This draws the conclusion that presenting similar users' ratings could be persuasive.

Subjective Evaluation. We now present the results of subjective evaluation by a Likert-scale questionnaire. Note that, in terms of efficiency, effectiveness and persuasiveness, they are also evaluated by directly asking participants. Table 3 presents the results of participants-reported efficiency, effectiveness, persuasiveness, satisfaction, trust and transparency. In subjective evaluation, *Simi* style explanations are perceived to be the most efficient, effective and persuasive. However, in objective evaluation (see Table 2), this is not the case. In objective evaluation, *Avg* style explanations are measured to be more efficient and more persuasive; *Context-aware* and *Content* style explanations are measured to be more effective. This shows that the actual effects of explanations may not be the same as those perceived by participants. How to eliminate the differences worth further investigation.

Satisfaction, trust and transparency are evaluated by directly asking participants. The last three columns in Table 3 contain the results of participants-reported satisfaction, trust and transparency. *Simi* style explanations achieve

Table 3. Subjective evaluation of efficiency, effectiveness, persuasiveness, satisfaction, trust and transparency. Note that these are the results of the Likert-scale questionnaire.

	Efficiency	Effectiveness	Persuasiveness	Satisfaction	Trust	Transparency
Avg	3.16	2.98	3.01	3.00	2.85	3.26
Content	3.25	3.12	3.20	3.19	3.28	**3.46**
Context-aware	2.96	3.07	3.24	3.21	3.14	3.26
Per	3.32	3.26	3.18	3.11	3.04	3.12
Simi	**3.72**	**3.47**	**3.52**	**3.50**	**3.53**	3.42
Simu	3.49	3.33	3.30	3.36	3.30	3.14

the highest level of satisfaction and trust, which shows that participants prefer explanations that compare the present recommendation and the items that they consumed before. *Content* style explanations provide information about recommendations and helps participants understand how the system works. As a result, *Content* style explanations are perceived to be transparent.

Correlation Between Different Metrics. Table 4 presents the *Spearman Rank Order Correlation* of metrics for evaluating *Context-aware* explanations, it can be observed that these metrics are not independent. For example, in the *Context-aware* explanation, trust and satisfaction, efficiency and effectiveness are positively correlated. However, transparency and persuasiveness, trust and efficiency are negatively correlated, this leads to the conclusion that it is hard to design explanations that can perform well in all aspects. In order to comprehensively evaluate the qualities of explanations, more metrics should be considered.

Table 4. Spearman Rank Order Correlation of metrics in *Context-aware* explanations.

	Efficiency	Effectiveness	Persuasiveness	Satisfaction	Trust	Transparency
Efficiency	1.0***	0.45***	0.05	−0.17	−0.15	0.01
Effectiveness	0.45***	1.0***	0.01	−0.08	-0.0	−0.1
Persuasiveness	0.05	0.01	1.0***	0.04	−0.03	−0.17
Satisfaction	−0.17	−0.08	0.04	1.0***	0.82***	−0.04
Trust	−0.15	−0.0	−0.03	0.82***	1.0***	−0.11
Transparency	0.01	−0.1	−0.17	−0.04	−0.11	1.0***

To conclude, the major findings from the results of our experiments are: (1) *Context-aware* explanations can help users better estimate the qualities of recommendations, however it takes longer time for users to make a decision; (2) Present similar users' ratings to explain recommendations could be persuasive; (3) The actual effects of explanations may not be the same as those perceived by participants; (4) The goals of explanations correlate with each other.

5 Conclusions and Perspectives

We now conclude and propose future avenues. The main contributions of this work are: (1) We develop a web-based questionnaire that can generate and explain recommendations; (2) We compare context-aware and non-contextual explanations in terms of transparency, effectiveness, efficiency, persuasiveness, trust and satisfaction with this questionnaire.

The major conclusions of our user study are: (1) Contrary to our hypothesis, contextual information alone may not boost the expected advantages of explanation in terms of transparency, efficiency, persuasiveness, trust and satisfaction, however *Context-aware* explanations helps users better estimate the quality of recommendations; (2) We point out that human perceived efficiency, effectiveness, persuasiveness might differ from the actual measurements. Further research in cognitive science is needed to explain this observation to better guide the design of explanation in RSs; (3) We find that metrics of explanations are not independent, more metrics should be considered to evaluate the quality of explanations.

The limitations of our work are: (1) In this work, we limited our work in movie recommendation; (2) The situation we used to generate recommendation is not the participants' real situation; (3) We explored explanations containing only contextual information; (4) The participants of the questionnaire are mostly students or employee of higher education. However, we believe that this work is an important step in exploring the effects of contextual information in explanations in recommender systems.

We now present perspectives for future work: (1) Extend our work to other application domains, such as restaurant recommendations, tour recommendations, etc.; (2) We plan to extend our experiment to a wider range of population in order get more reliable conclusions; (3) Combine contextual information and other information such as content and demographic information to serve as explanations and explore how they can influence the effects of explanations.

References

1. Adomavicius, G., Tuzhilin, A.: Context-Aware Recommender Systems. In: Ricci, F., Rokach, L., Shapira, B., Kantor, P. (eds.) Recommender Systems Handbook, pp. 217–253. Springer, Boston, MA (2011). https://doi.org/10.1007/978-0-387-85820-3_7
2. Abowd, G.D., Dey, A.K., Brown, P.J., Davies, N., Smith, M., Steggles, P.: Towards a better understanding of context and context-awareness. In: Gellersen, H.W. (ed.) Handheld and Ubiquitous Computing. HUC 1999. Lecture Notes in Computer Science, vol. 1707, pp. 304–307. Springer, Heidelberg (1999). https://doi.org/10.1007/3-540-48157-5_29
3. Haruna, K., et al.: Context-aware recommender system: A review of recent developmental process and future research direction. Appl. Sci. **7**(12), 1211 (2017)
4. Tintarev, N., Masthoff, J.: Explaining recommendations: Design and evaluation. In: Ricci, F., Rokach, L., Shapira, B. (eds.) Recommender Systems Handbook, pp. 353–382. Springer, Boston, MA (2015). https://doi.org/10.1007/978-1-4899-7637-6_10

5. Gedikli, F., Jannach, D., Ge, M.: How should i explain? a comparison of different explanation types for recommender systems. Int. J. Human Comput. Stud. **72**(4), 367–382 (2014)
6. Zhang, Y., Chen, X.: Explainable recommendation: a survey and new perspectives. arXiv preprint: arXiv:1804.11192 (2018)
7. Musto, C., Spillo, G., de Gemmis, M., Lops, P., Semeraro, G.: Exploiting distributional semantics models for natural language context-aware justifications for recommender systems. In: IntRS@ RecSys, pp. 65–71 (2020)
8. Baltrunas, L., Ludwig, B., Peer, S., Ricci, F.: Context-aware places of interest recommendations and explanations. In: Joint Proceedings of the Workshop on Decision Making and Recommendation Acceptance Issues in Recommender Systems (DEMRA 2011) and the 2nd Workshop on User Models for Motivational Systems: The Affective and the Rational Routes to Persuasion (UMMS 2011), CEUR Workshop Proceedings. vol. 740, pp. 19–26 (2011)
9. Misztal, J., Indurkhya, B.: Explaining contextual recommendations: interaction design study and prototype implementation. In: IntRS@ RecSys, pp. 13–20 (2015)
10. Sato, M., Ahsan, B., Nagatani, K., Sonoda, T., Zhang, Q., Ohkuma, T.: Explaining recommendations using contexts. In: 23rd International Conference on Intelligent User Interfaces, pp. 659–664 (2018)
11. Zhong, J., Negre, E.: Towards better representation of context into recommender systems. In: IJKBO (2022, to appear)
12. Ferdousi, Z.V., Colazzo, D., Negre, E.: CBPF: leveraging context and content information for better recommendations. In: Gan, G., Li, B., Li, X., Wang, S. (eds.) Advanced Data Mining and Applications. ADMA 2018. Lecture Notes in Computer Science, vol. 11323, pp. 381–391. Springer, Cham (2018). https://doi.org/10.1007/978-3-030-05090-0_32
13. Benesty, J., Chen, J., Huang, Y., Cohen, I.: Pearson correlation coefficient. In: Benesty, J., Chen, J., Huang, J., Cohen, I. (eds.) Noise Reduction in Speech Processing. Springer Topics in Signal Processing, vol. 2, pp. 1–4. Springer, Heidelberg (2009). https://doi.org/10.1007/978-3-642-00296-0_5
14. Košir, A., Odic, A., Kunaver, M., Tkalcic, M., Tasic, J.F.: Database for contextual personalization. Elektrotehniški vestnik **78**(5), 270–274 (2011)
15. Ferdousi, Z.V.: From Traditional to Context-Aware Recommendations by Correlation-Based Context Model. Ph.D. thesis, Paris-Dauphine University, PSL Research University (2020)
16. Bilgic, M., Mooney, R.J.: Explaining recommendations: Satisfaction vs. promotion. In: Beyond Personalization Workshop, IUI, vol. 5, p. 153 (2005)
17. Herlocker, J.L., Konstan, J.A., Riedl, J.: Explaining collaborative filtering recommendations. In: Proceedings of the 2000 ACM Conference on Computer Supported Cooperative Work, pp. 241–250 (2000)
18. Sarwar, B., Karypis, G., Konstan, J., Riedl, J.: Item-based collaborative filtering recommendation algorithms. In: Proceedings of the 10th International Conference on World Wide Web, pp. 285–295 (2001)
19. Vig, J., Sen, S., Riedl, J.: Tagsplanations: explaining recommendations using tags. In: Proceedings of the 14th International Conference on Intelligent User Interfaces, pp. 47–56 (2009)
20. Tabachnick, B.G., Fidell, L.S.: Experimental Designs Using ANOVA. Thomson/Brooks/Cole Belmont, CA (2007)
21. Abdi, H., Williams, L.J.: Tukey's honestly significant difference (HSD) test. Encycl. Res. Des. **3**(1), 1–5 (2010)

Improved Local Binary Pattern for Face Recognition

Shekhar Karanwal(✉)

CSE Department, Graphic Era University (Deemed), Dehradun 248002, Uttarakhand, India
shekhar.karanwal@gmail.com

Abstract. This work gives rise to novel LBP variant so-called Improved LBP (ILBP) for Face Recognition (FR), to overcome the LBP limitation. LBP unable to produce the balanced transformed image for histogram extraction by utilizing only the neighbors gray pixel comparison with the center pixel. ILBP launches robust methodology in 2 steps. First there is computation of gray level difference between neighbors and its mean and second negative threshold is used for comparison with gray level difference. By utilizing mean and negative threshold there is evolution of ILBP transformed image which is more balanced than LBP. As 2 negative thresholds are used therefore 2 ILBP transformed images are produced. The histograms (region wise) of ILBP transformed image is so impressive that it beats histograms of LBP, HELBP, MBP and HOG. ILBP achieves best recognition rates of 99.58% and 90.85% on ORL and GT datasets, which is much higher than the 4 compared descriptors. Several literature based methods are also outclassed by ILBP. PCA and SVMs are used for compression.

Keywords: Local Binary Pattern (LBP) · Improved LBP (ILBP) · Horizontal Elliptical LBP (HELBP) · Other approaches

1 Introduction

The substantial attention has been attained by local descriptors from last 3 decades. Their discriminative in unfavorable conditions is quite appreciable. In local descriptors, the features of distinct local parts (of input image or transformed image) are merged for complete feature size making. The distinct local parts are eyes, mouth, nose and lips. Among several local descriptors invented in literature, LBP [1] is most promising one. LBP simplicity is due to non-complex algorithm and ease of evaluation. After LBP appearance, there is the contribution of several LBP variants in literature.

Raghuwanshi et al. [2] presented texture based descriptor Hybrid Directional Extrema Pattern (HDEP). By integrating WD_DLEP and DLEP features there is feature production of the proposed descriptor. The texture patterns are evaluated from 4 directions. The launched method considers difference among center and neighbor pixels in defined directions. Impressive results are attained by HDEP method. Karanwal et al. [3] evaluated 14 local descriptors (for comparison) in FR. The feature extraction is done in

L. Troiano et al. (Eds.): ICDLAIR 2021, LNNS 441, pp. 86–96, 2022.
https://doi.org/10.1007/978-3-030-98531-8_9

global way and their sizes are compressed by PCA and FLDA. On numerous unfavorable challenges it is CLBP that accomplish best results on most times. MRELBP-NI also attains good outcomes. Shamaileh et al. [4] presented Feat-WCLTP descriptor for texture investigation. In Feat-WCLTP, there is merging of RDWT with CLTP and then variance and mean features are utilized for describing magnitude details. Various descriptors are conquered by the Feat-WCLTP. Karanwal et al. [5] invented 2 color descriptors for FR called CZZBP and CMBZZBP. The functionality of 2 launched descriptors are based on zigzag pixels alignment. In CZZBP, the pattern length is build by the subtraction of IInd order pixel from the Ist order pixel and in CMBZZBP the same is carried out on median patch. Both pulls amazing outcomes. Song et al. [6] proposed 2 novel descriptors Local Grouped Order Pattern (LGOP) and Non LBP (NLBP) for textures. In LGOP, by considering dominant direction, neighbor sampling locations are grouped and relationship of intensity order is encoded group wise. In NLBP, numerous anchors are computed based on statistics of global image and gradually intensity differences (non-local) are encoded among anchors and neighboring locations. Further LGOP and NLBP are fused to form the robust feature LGONBP. LGONBP beats many descriptors results. Karanwal et al. [7] found the discriminative descriptor for FR called NCDB-LBP. The launching of 4 labeled function, for making the NCDB-LBP feature size is its prime strength. In both directions (i.e. clockwise and anticlockwise) excellent outcomes are attained by NCDB-LBP. Sun et al. [8] introduced LGNP for FR. Precisely, the sobel operator is employed initially for production of gradient image. Then LGNP code is derived based on LDP methods. Further regional and global information are fused by FCCP, named as FCCP_LGNP, which outclasses results of many others. Karanwal et al. [9] discovered the OD-LBP descriptor in numerous challenging environments. The OD-LBP size formation is based on three differences (gray) determined for each and every orthogonal location. The launched method proves its excellence on numerous datasets. Raghavendra et al. [10] gives novel descriptor for face anti-spoofing ELTCP. In ELTCP, the ternary edges co-relation is encoded by using center pixel in conjunction of its immediate and next immediate neighbors. The ELTCP produces better outcomes than several other ones. Suruliandi et al. [11] develops the texture based descriptor Local Symmetric Tetra Pattern (LSTP). For describing local regions of the image the descriptor is developed. In addition to the 8 surrounding neighbors, the proposed LSTP also considers pixels (8) at next level for describing texture more efficiently. Promising consequences are attained by LSTP on numerous datasets. Some more details of the LBP variants are found in [12–15].

Problem Formulation: Most of local descriptors suffers somewhere in the robustness due to their incomplete methodologies. There are very less descriptors exist in literature which achieve excellence by using the 8 neighborhoods structure. There is need for such descriptor which should be discriminant and based on 8 neighbor structure.

Novelty: The proposed work introduces the novel descriptor for FR by using 8 neighborhood structure in 3×3 patch. This novel descriptor is Improved LBP (ILBP).

Contributions: This work gives rise to novel LBP variant Improved LBP (ILBP) for FR, to overcome the LBP limitation. LBP unable to produce balanced transformed image for histogram extraction by utilizing only neighbors gray pixel comparison with center pixel. ILBP launches robust methodology in 2 steps. First there is computation of gray level difference between neighbors and its mean and second negative threshold is used for comparison with gray level difference. By using mean and negative threshold there is evolution of ILBP transformed image which is more balanced than LBP. As 2 negative thresholds are used therefore 2 ILBP transformed images are produced. The histograms (region wise) of ILBP transformed image is so impressive that it beats histograms of LBP [1], HELBP [16], MBP [17] and HOG [18]. Several literature methods are also outclassed by ILBP. PCA [19] and SVMs [20] are used for compression and matching. Evaluation is done on ORL [21] and GT [22] datasets.

Road Map: The organization of the rest of the paper is given as: LBP, HELBP, MBP and HOG descriptors are debated in Sect. 2, the proposed ILBP with full FR framework are discussed in Sect. 3, experiments are portrayed in Sect. 4 and finally conclusions with future scope are placed in Sect. 5.

2 Illustration of LBP, HELBP, MBP and HOG Descriptors

2.1 Local Binary Pattern (LBP)

LBP was launched by Ojala in [1] for textures. In LBP, the gray pixel intensity of the 8 neighbors are compare with center gray pixel intensity in 3 × 3 patch. 1 is set to those neighbor places who have big or similar value to center gray pixel, else 0 is set. From these places the binary pattern of 8 bit is formed, whose conversion in decimals is done by weights granting. After obtaining decimal codes of each location, there is formation of LBP image. From LBP image, 3 × 3 LBP histograms (in regional way) are extracted and merged, which comes up with total size of 2304 (as 256 is 1 region size). Equation 1 define LBP concept. In Eq. 1, A and B designate neighbors size and radius. $V_{B,a}$ and V_c are separate neighbor values and center value.

$$LBP_{A,B}(x_c) = \sum_{a=0}^{A-1} k(V_{B,a} - V_c)2^a,\ k(y) = \begin{pmatrix} 1\ y \geq 0 \\ 0\ y < 0 \end{pmatrix} \tag{1}$$

2.2 Horizontal Elliptical Local Binary Pattern (HELBP)

Nguyen et al. [16] explore HELBP concept in FR. In HELBP, the horizontal gray pixel intensity of the 8 neighbors are compare with the center gray pixel intensity in 3 × 5 patch. 1 is set to those neighbor places who have big or similar value to center gray pixel, else 0 is set. This leads to binary pattern of 8 bit, whose conversion in decimals is done by the weights granting. After obtaining decimal codes of each location, there is formation of HELBP image. From HELBP image, 3 × 3 HELBP histograms (regionally) are extracted and merged, which builds total size of 2304 (256 is 1 region size). Equation 2 defines HELBP concept. In Eq. 2, A, B_1 and B_2 designate neighbors size and radius. $V_{B_1,B_2,a}$ and V_c are separate neighbor values and center value.

$$HELBP_{A,B_1,B_2}(x_c) = \sum_{a=0}^{A-1} k(V_{B_1,B_2,a} - V_c)2^a, k(y) = \begin{pmatrix} 1\ y \geq 0 \\ 0\ y < 0 \end{pmatrix} \tag{2}$$

2.3 Median Binary Pattern (MBP)

Hafiane et al. [17] launched MBP descriptor for textures. In MBP, the gray pixel intensity of the 9 pixels are compare with its median, in 3 × 3 patch. 1 is set to those places who have big or similar value to median value, else 0 is set. From these places the binary pattern of 9 bit is formed. The center bit is optional in MBP, to include or not. For this work it is not included which form the 8 bit pattern. Then conversion in decimals is done by weights granting. After obtaining decimal codes of each location, there is formation of MBP image. From MBP image, 3 × 3 MBP histograms (in regional way) are extracted and merged, which comes up with total size of 2304 (as 256 is 1 region size). Equation 3 and Eq. 4 states MBP concept. In Eq. 3, A designate size of patch and in Eq. 4, A designate neighbors size. B, $V_{B,a}$ and V_{median} are the radius, separate values and median value.

$$MBP_{A,B}(x_c) = \sum_{a=0}^{A-1} k(V_{B,a} - V_{median})2^a,\ k(y) = \begin{cases} 1 & y \geq 0 \\ 0 & y < 0 \end{cases} \text{ for } A = 9 \quad (3)$$

$$MBP_{A,B}(x_c) = \sum_{a=0}^{A-1} k(V_{B,a} - V_{median})2^a,\ k(y) = \begin{cases} 1 & y \geq 0 \\ 0 & y < 0 \end{cases} \text{ for } A = 8 \quad (4)$$

2.4 Histogram of Oriented Gradients (HOG)

For human detection, Dalal et al. [18] launched its descriptor, named as HOG. By doing the convolution between image regions and 1-D mask [−1 0 1] at $\sigma = 0$, the gradients images (from regions) are produced. The histograms of these gradients are assembled in the 9 bin size. The recognition rate falls down if the bin size of above 9 is used. Then overlapping and contrast enhancement is carried in p × q blocks. There is improvement in recognition rates after overlapping. For the enhancement of contrast, 4 methods can be considered. These 4 ones are (L2-L1) norm, L1-sqrt and L2-hys. In this work, the features of HOG is carried out from 9 regions of 3 × 3 patch. The content of overlapping utilized is 50% by using the 2 × 2 block size, so full HOG size is 144.

3 Description of the Proposed Descriptor and Full FR Framework

3.1 The Proposed Descriptor Improved Local Binary Pattern (ILBP)

Improved LBP (ILBP) is the novel LBP variant developed for Face Recognition (FR), to overcome the LBP limitation. LBP unable to produce the balanced transformed image for histogram extraction by utilizing only neighbors gray pixel comparison with center pixel. In ILBP, initially there is computation of the gray level difference between neighbors and its mean. Then negative threshold is used for the comparison with the gray level difference. 1 is set to those places who (gray differences) have big or similar value to the specified negative threshold, else 0 is set. As 2 negative thresholds are considered therefore from these places 2 binary pattern of 8 bits are formed, whose conversion in decimals is done by weights granting. After obtaining decimal codes of each location, there is formation of the 2 ILBP images at **T = −1 and T = −2**. From both ILBP images,

3 × 3 ILBP histograms (in regional way) are extracted and merged, which forms **ILBP** size of 2304 (as 256 is 1 region size). The histogram extracted from **ILBP** image is more discriminant than LBP (as proved experimentally) and also from many others. Equation 5 define ILBP concept for both thresholds. In Eq. 5, A, B and T designate the neighbors size, radius and threshold. $V_{B,a}$ and $\mu_{B,A}$ are the separate neighbor values and mean value. In Eq. 6, the mean computation is expressed. Figure 1 and Fig. 2 describe ILBP illustration at $T = -1$ and $T = -2$. Figure 3 and Fig. 4 gives histogram representation.

$$ILBP_{A,B,T}(x_c) = \sum_{a=0}^{A-1} k(V_{B,a} - \mu_{B,A})2^a,\ k(y) = \begin{cases} 1 & \text{for } y \geq T \\ 0 & \text{for } y < T \end{cases} \quad (5)$$

$$\mu_{B,A} = \frac{1}{A} \sum_{a=0}^{A-1} V_{B,a} \quad (6)$$

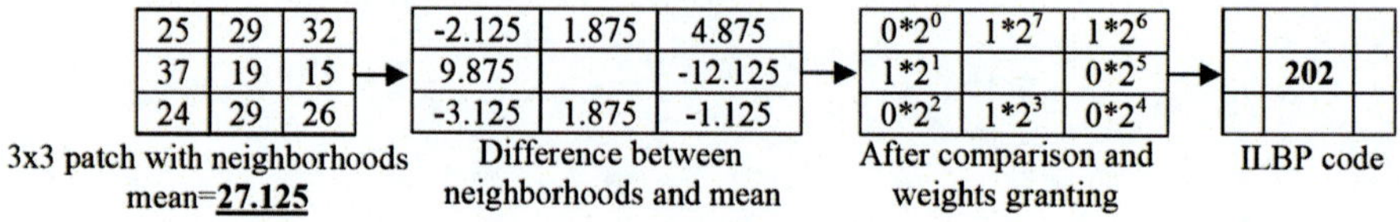

Fig. 1. Demonstration of ILBP descriptor at $T = -1$

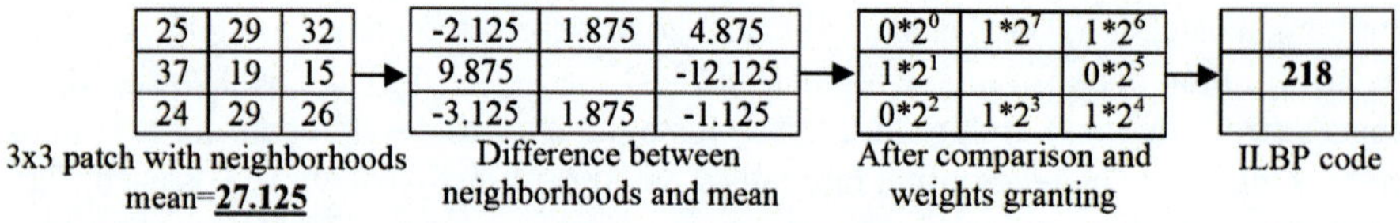

Fig. 2. Demonstration of ILBP descriptor at $T = -2$

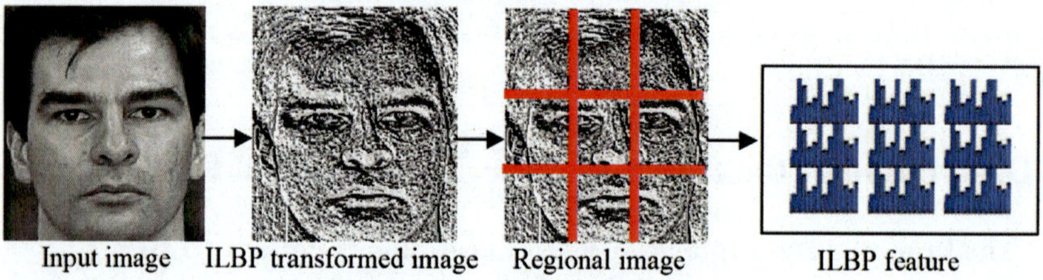

Fig. 3. ILBP image with regional histogram at $T = -1$

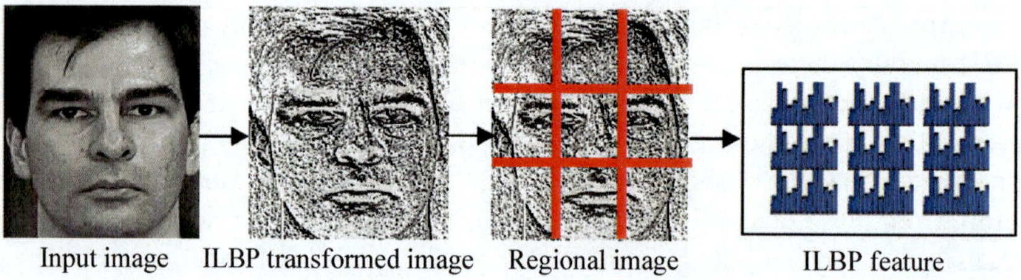

Fig. 4. ILBP image with regional histogram at $T = -2$

3.2 Full FR Framework

As features are extracted regionally therefore ILBP builds the large feature size. It is required to cut down the dimension and chose the relevant features for classification. This task of dimension cutting is performed by PCA. For classification Radial Basis Function (RBF) is deployed, which is SVMs based classifier. The block diagram of the proposed FR framework is displayed in Fig. 5 and Fig. 6. Before deploying feature extraction step, the images are resized so that computational cost is reduced. To establish fair comparison among introduced method and the other compared ones, the tasks of resizing, feature extraction, dimension reduction and classification also performed for the compared descriptors. Rest details are given in experiments section.

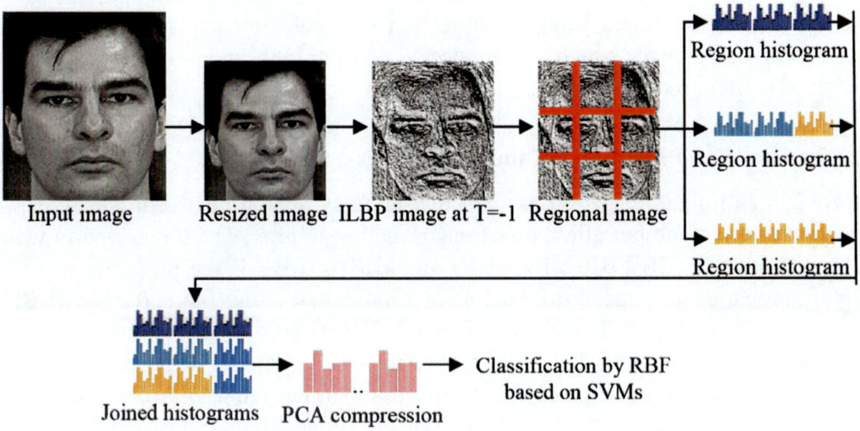

Fig. 5. Proposed FR framework at T = −1

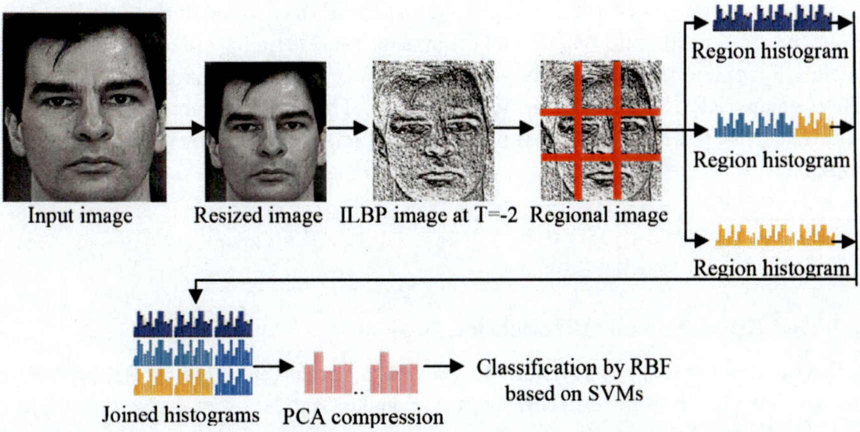

Fig. 6. Proposed FR framework at T = −2

4 Experiments

4.1 Dataset Details

The two datasets details which are used for the evaluation are given below.

4.1.1 ORL Dataset

The ORL dataset is created from 40 subjects. Every subject carry 10 distinct images acquired in 3 challenges which are pose, emotion and lightning. The total count is 400 and all have the size of 112 × 92.

4.1.2 GT Dataset

The GT dataset is created from 50 subjects. Every subject carry 15 distinct images acquired in 4 challenges which are emotion, lightning, scale and pose. The total count are 750. The image size are varying in GT dataset.

4.2 Feature Size Particulars of the Descriptors

On ORL, the original image resolution is resized to 49 × 46 for extracting features. On GT, the same is done but after color transformation to gray. The 5 descriptors which are tested are LBP, HELBP, MBP, HOG and **ILBP**. **ILBP** is the proposed one. The size of the features produced from them are 2304, 2304, 2304, 144 and **2304**. ILBP is tested at two T values i.e. $\mathbf{T = -1}$ and $\mathbf{T = -2}$. PCA is the holistic method adopted for compressing the size. After PCA, 29 and 40 are sizes exploited by the RBF classifier on ORL and GT. All descriptors are evaluated on similar reduced dimensions so that comparison is accomplished fairly. Simulation is conducted in MATLAB R2018a.

4.3 Estimation of Recognition Rate

The recognition rate (%) estimation is based on formula as defined in Eq. 7. In Eq. 7, RR denote the recognition rate, M_{ts} denote the size of test set and I_{fm} are the incorrect found matches. For example if I_{fm} have the value 4 and M_{ts} is 360 then RR is 356/360 = 98.88%. Which means RR is examined on test set (M_{ts}). The other variable M_{trs} signify the training set. The partition between training and test sets are formed by using the 'holdout' approach.

$$[RR = \frac{M_{ts} - I_{fm}}{M_{ts}} * 100] \tag{7}$$

4.3.1 RR Estimation on ORL Dataset

On this dataset, the ($\mathbf{M_{trs} = 1{:}4}$) and ($\mathbf{M_{ts} = 9{:}6}$). The classifier topmost RR is observed after 40 runs. The introduced **ILBP** at both T values outclass remaining 4 descriptors. Table 1 deliver all the results. Table 1 clearly shows the dominance of **ILBP** than the other compared descriptors on both T values. The percentage difference between ILBP and these descriptors are between (4.72–11.11) percent, which is quite astonishing. ILBP defeats all these compared descriptors by very large margin.

4.3.2 RR Estimation on GT Dataset

On this dataset, the **(M_{trs} = 5:8)** and **(M_{ts} = 10:7)**. The classifier topmost RR is observed after 37 runs. The introduced **ILBP** at both T values outclass remaining 4 descriptors. Table 1 deliver all the results. Table 1 clearly shows the dominance of **ILBP** than the other compared descriptors on both T values. The percentage difference between ILBP and these descriptors are between (3.00–11.20) percent, which is quite remarkable. ILBP defeats all these compared descriptors by very large margin.

Table 1. RR evaluation on ORL and GT datasets

	On ORL				On GT			
	M_{trs} values				M_{trs} values			
	$M_{trs} = 1$	$M_{trs} = 2$	$M_{trs} = 3$	$M_{trs} = 4$	$M_{trs} = 5$	$M_{trs} = 6$	$M_{trs} = 7$	$M_{trs} = 8$
Descriptors		RR in %				RR in %		
LBP	78.61	90.93	96.07	98.75	82.20	83.55	85.75	87.14
HELBP	77.22	90.31	96.07	97.50	79.80	83.33	85.50	86.00
MBP	77.22	90.31	96.07	97.50	80.60	83.55	85.50	86.00
HOG	72.22	85.93	92.85	96.25	74.00	77.33	79.25	82.85
ILBP (T = −1)	**82.50**	**94.06**	**98.57**	**99.16**	**84.00**	**85.33**	**88.50**	**89.14**
ILBP (T = −2)	**83.33**	**94.37**	**97.85**	**99.58**	**85.20**	**86.66**	**89.75**	**90.85**

4.4 Comparison of RR with Literature Methods

As **ILBP at T = −2** succeeds in obtaining better results than **ILBP at T = −1** so **ILBP at T = −2** is compared with methods from literature. The detailed investigation of RR on ORL and GT datasets are given below. Various literature methods are picked from literature and used for comparison. These methods are local and non-local based.

4.4.1 Comparison on ORL Dataset

On ORL, total amount of the methods compared are 9. ILBP fully conquers the RR of 9 other methods. Table 2 deliver all RR. These compared methods falls into category of dimension reduction, local and sparse. It is observed from Table 2 that in most of M_{trs} values there is huge difference between RR of ILBP and compared methods.

4.4.2 Comparison on GT Dataset

On GT also, total amount of methods compared are 9. ILBP again fully outclass the RR of 9 other methods. Table 3 deliver all the RR. These compared methods falls into the category of dimension reduction and local. It is witnessed from Table 3 that in most of M_{trs} values there is huge difference between RR of ILBP and compared methods.

Table 2. RR comparison on ORL dataset

	M_{trs} values			
	$M_{trs} = 1$	$M_{trs} = 2$	$M_{trs} = 3$	$M_{trs} = 4$
All methods	RR in %			
WNPEE [23]	N/A	N/A	N/A	84.80
NPE [23]	N/A	N/A	N/A	82.05
NCDB-LBPac [7]	80.27	93.75	97.14	99.16
NCDB-LBPc [7]	81.11	92.50	96.78	98.75
OD-LBP [9]	79.44	92.50	97.14	N/A
CLBP [3]	73.05	87.81	94.28	95.00
ELBP [3]	67.77	84.68	90.71	93.75
SGE [24]	N/A	91.75	93.14	95.14
LS-KSRC [25]	N/A	N/A	89.64	92.75
ILBP (T = −2)	**83.33**	**94.37**	**97.85**	**99.58**
N/A-Not Available				

Table 3. RR comparison on GT dataset

	M_{trs} values			
	$M_{trs} = 5$	$M_{trs} = 6$	$M_{trs} = 7$	$M_{trs} = 8$
All methods	RR in %			
WNPEE [23]	63.84	N/A	69.83	N/A
NPE [23]	56.65	N/A	59.59	N/A
NCDB-LBPac [7]	84.40	86.22	88.50	90.28
NCDB-LBPc [7]	84.00	85.55	89.25	90.85
CLBP [3]	N/A	N/A	87.25	89.14
RD-LBP [3]	N/A	N/A	77.50	80.28
CZZBP [5]	N/A	81.33	82.75	85.14
CMBZZBP [5]	N/A	84.44	87.75	88.57
LDV [26]	65.92	N/A	N/A	N/A
ILBP (T = −2)	**85.20**	**86.66**	**89.75**	**90.85**
N/A-Not Available				

5 Conclusions with Future Scope

This work presented the novel descriptor for FR so-called ILBP. ILBP adopted the total different methodology than LBP and many other LBP variants. ILBP produces more balanced transformed image than LBP as a result the robust histogram size feature is build. In ILBP, first there is computation of the gray level difference between neighbors and its mean and second the negative threshold is used for the comparison with the gray level difference. The reduction in size is pulls off by the usage of PCA and matching is carried out by RBF (SVMs based). For testing, ORL and GT datasets are considered. The ILBP descriptor achieves amazing outcomes on both datasets as a result it fully conquer the results of 22 methods. Among 22, the 4 descriptors i.e. LBP, HELBP, MBP and HOG are executed along with proposed ILBP and then tested. The remaining 18 are those methods which are literature based. By defeating the recognition rates of numerous methods, the ILBP justify its relevance on 2 datasets.

Some of limitations which are detected in the proposed method are: First, there is absence of multiscale feature extraction, which can add more robustness proposed feature. Second, testing of proposed method on other challenges such as noise, blur and occlusion changes. Third, to test the descriptor on different applications, which makes the proposed descriptor more versatile descriptor. All these crucial steps will be appended in the futuristic research. In addition, the novel color based descriptor will be introduced in the upcoming research.

References

1. Ojala, T., Pietikainen, M., Harwood, D.: A comparative study of texture measures with classification based on featured distributions. Pattern Recognit. **29**(1), 51–59 (1996)
2. Raghuwanshi, G., Tyagi, V.: Texture image retrieval using HDEP. Multimedia Tools Appl. **80**, 2295–2317 (2021)
3. Karanwal, S.: A comparative study of 14 state of art descriptors for face recognition. Multimedia Tools Appl. **80**(8), 12195–12234 (2021). https://doi.org/10.1007/s11042-020-09833-2
4. Shamaileh, A.M., Rassem, T.H., Chuin, L.S., Sayaydeh, O.N.A.: A new feat-WCLTP for texture image classification. IEEE Access **8**, 28276–28288 (2020)
5. Karanwal, S., Diwakar, M.: Two novel color local descriptors for face recognition. Optik **226**(2), 1–15 (2021)
6. Song, T., Feng, J., Luo, L., Gao, C., Li, H.: Robust texture description using LGOP and NLBP. IEEE Trans. Circuit Syst. Video Technol. **31**(1), 189–202 (2021)
7. Karanwal, S., Diwakar, M.: Neighborhood and center difference based LBP for face recognition. Pattern Anal. Appl. **24**(2–3), 741–761 (2021)
8. Sun, J., Lv, Y., Tang, C., Sima, H., Wu, X.: Face recognition based on LGNP and FCCP. IEEE Access **8**, 35777–35791 (2020)
9. Karanwal, S., Diwakar, M.: OD-LBP: orthogonal difference-LBP for face recognition. Digit. Signal Process. **110**(3), 102948 (2021)
10. Raghavendra, R.J., Kunte, R.S.: ELTCP: a novel feature descriptor for face anti-spoofing. J. Inf. Secur. Appl. **52**, 1–10 (2020)
11. Suruliandi, A., Sinduja, A., Raja, S.P.: Texture classification using the rotational-invariant LSTP. Int. J. Wavlets Multisolut. Inf. Process. **17**(4), 1–23 (2019)
12. Kumar, D., Garain, J., Kisku, D.R., Singh, J.K., Gupta, P.: Unconstrained and constrained face recognition using DLD with ensemble framework. Neuro **408**, 273284 (2020)
13. khadiri, I.E., Chahi, A., Merabet, Y.E., Ruichek, Y., Touahni, R.: Image classification with LDDTP. In: ICCDIT, pp. 812–817 (2019)

14. Acharya, S., Nanda, P.K.: Adjacent LBP and LTP based background modeling with mixed-mode learning for foreground detection. Pattern Anal. Appl. **24**(3), 1047–1074 (2021). https://doi.org/10.1007/s10044-021-00967-z
15. Chen, W.S., Jeng, R.H.: A new patch-based LBP with adaptive weights for gender classification of human face. J. Chin Inst. Eng. **43**(5), 451–457 (2020)
16. Nguyen, H.-T., Caplier, A.: Elliptical local binary patterns for face recognition. In: Park, J.-I., Kim, J. (eds.) ACCV 2012. LNCS, vol. 7728, pp. 85–96. Springer, Heidelberg (2013). https://doi.org/10.1007/978-3-642-37410-4_8
17. Hafiane, A., Seetharaman, G., Zavidovique, B.: Median binary pattern for textures classification. In: Kamel, M., Campilho, A. (eds.) ICIAR 2007. LNCS, vol. 4633, pp. 387–398. Springer, Heidelberg (2007). https://doi.org/10.1007/978-3-540-74260-9_35
18. Dalal, N., Triggs, B.: HOG for human detection. In: ICVPR, pp. 886–893 (2005)
19. Jia, Y., Liu, H., Hou, J., Kwong, S., Zhang, Q.: Semisupervised affinity matrix learning via dual-channel information recovery. IEEE Trans. Cybern. 1–12 (2021)
20. Ma, Z., Li, B.: A DDoS attack detection method based on SVM and K-nearest neighbour in SDN environment. Int. J. Comput. Sci. Eng. **23**(3), 224–234 (2020)
21. http://www.cl.cam.ac.uk/research/dtg/attarchive/facedatabase.html
22. http://www.anefian.com/research/face_reco.htm
23. Mehta, S., Zhan, B.S., Shen, X.J.: WNPEE. Electrronics **8**(2), 1–18 (2019)
24. Wan, M., Ge, M., Zhan, T., Yang, Z., Zheng, H., Yang, G.: SGE based on the fuzzy set for image classification. Computer 1–10 (2021)
25. Tan, S., Sun, X., Chan, W., Qu, L., Shao, L.: Robust face recognition with KL-SGSR. IEEE Trans. Image Process. **26**(10), 4661–4668 (2017)
26. Wu, X., Sun, J.: Face recognition based on multi-scale LDV. Multimedia Tools Appl. **79**, 2409–2425 (2020)

Incorporating Dynamic Information into Content-Based Recommender System in Online Learning Environment

Qing Tang[1(✉)], Marie-Hélène Abel[1], and Elsa Negre[2]

[1] Sorbonne Universités, Université de Technologie de Compiègne, CNRS UMR 7253, HEUDIASYC, 60203 Compiègne, France
{qing.tang,marie-helene.abel}@utc.fr

[2] Paris-Dauphine University, PSL Research University, CNRS UMR 7243, LAMSADE, 75016 Paris, France
elsa.negre@dauphine.fr

Abstract. Learners' information (e.g., preferences, competencies, contexts) changes over time, which makes it difficult for the recommender system to accurately predict learners' current needs. Education has higher requirements for recommendation accuracy, especially in online learning environments, it is crucial to accurately understand learners' current needs and maintain their enthusiasm for learning. We propose an approach to incorporate learners' dynamic information (preferences, competencies, and contexts) into a content-based recommender system to help learners acquire apposite resources. The system calculates learners' preferences and competencies from recent learning activities, at the same time, establishes connections between context, concentration, and resource types, makes recommendations more explainable and adaptive.

Keywords: Online learning · Dynamic information · Recommender system

1 Introduction

Nowadays, the massive quantity of information on the internet makes information-seeking a difficult task, both information providers and consumers are facing challenges: information providers want the information to be transferred to the target consumers while information consumers are desired to find the information most relevant to their needs [1]. Many Online Learning Environments (OLEs) are facing the similar problems, learners have information burden and difficulty in resource decision-making when facing with so many diverse resources [2]. Recommender Systems (RSs) are widely applied in various fields and help people to find information matching their desires [3]. We want to use RSs to help learners in OLEs efficiently obtain suitable educational resources for their current learning processes.

L. Troiano et al. (Eds.): ICDLAIR 2021, LNNS 441, pp. 97–106, 2022.
https://doi.org/10.1007/978-3-030-98531-8_10

Learners' information is changing over time [4]. Stationary RSs tend to only get learners' desires based on static information, which assume that learners' needs are constant [3]. However, preferences will change over time, which are highly subjective; a period of non-learning will lead to a decline of learners' competencies; a usage of appropriate resource will improve competencies; when the surroundings (contexts) change, learners' concentration will change accordingly, and different levels of concentration make learners choose different types of resources [5]. The above mentioned are the dynamic factors that will affect learners' choices of resources.

In this paper, we propose a Dynamic Content-based Recommender System (DCRS) to address the above issues. Specifically, we propose to extract learners' competency and preference information from activity records collected from the online learning platform. And we also consider the connections between context, concentration, and resource type. Finally, we incorporate all the dynamic factors into a content-based RS.

The remaining parts of this paper are organized as follows: Sect. 2 reviews the most relevant existing work. Section 3 introduces the details of the proposed DCRS framework. Next in Sect. 4, we discuss the proposed method. Finally, we conclude this paper in Sect. 5.

2 Related Work

2.1 Online Learning Environment (OLE)

Online learning is a web-based learning activity, which through relevant online learning environments, learners achieve timely, resource-sharing and zero-distance learning experiences and approaches [6].

Currently, OLEs can be divided into three categories [7]. The first is the traditional platform, it mainly distributes video and audio through television or Internet, and is widely used in remote places and areas with poor educational quality. The second is represented by Massive Open Online Course (MOOC) [8], which has diversified learning tools and rich resources, integrates technological research and application innovation, and uses efficient precision tests and extensively connected social networks to provide learners with learning materials anytime, anywhere. The third is small private open course, which is not an alternative to MOOCs, but rather the inheritance, perfection and transcendence of MOOCs [9]. It absorbs and inherits advanced ideas of MOOCs, such as high-quality resources, accurate tests, timely feedbacks, personalized recommendations, and so on. In response to the high dropout rate of MOOCs and the lack of immersive learning experience, it has changed the conditions for the application of the course and increased the participation rate of the learners and teachers [7].

2.2 Dynamic Information

Learner's profile contains a set of information about learning process, such as knowledge levels and skills, learning styles, preferences, competencies, schedules

[10]. A standard learner's profile contains static and dynamic information. Static information refers to personal generic information, which does not change in a short period of time, such as learner's name, gender, nationality, specialty, degree, profession title. Dynamic information includes own information such as learner's knowledge levels, learning styles, preferences, competencies; external information such as learner's surroundings, environments.

Chen and Mizoguchi designed a communication content ontology for learner model agent in multi-agent architecture [11], in the learner model they defined, a clear distinction is made between static and dynamic learner information. Vagale and Niedrite defined a learner model as an abstract representation of the system learner [12], in addition to the profile data (static generic information), some specific information is also included, such as psychological information, and domain-dependent information. Although the above works mentioned dynamic information, these works did not form a systematic application, from definition to collection to application, and did not reflect the importance of dynamic information in OLEs.

2.3 Recommender System (RS)

RSs aim to provide users with personalized product or service recommendations to handle the increasing information overload problem and improve customer relationship management [13]. Various RS techniques have been proposed, and the common algorithms mainly are Content Based (CB), Collaborative Filtering (CF), Knowledge Based (KB), and Hybrid approaches.

The task of RSs in learning domain is to recommend relevant learning materials to the learners and help them in decision making [14]. Chen et al. proposed a RS to recommend learning materials in an online learning platform and their results demonstrated significant improvement in performance [15]. Takano and Li proposed a RS for online learning by utilizing a feedback method that extracts student's preferences and web-browsing behaviors [16]. Thus, we want to expand the exploration of RS in OLE, and improve its performance.

3 Methodology

In this section, we first present the overall structure of the proposed DCRS framework. Then, enumerate its components (activity records, dynamic learner model, resource profile). Finally, we incorporate the components into a content-based recommender system, and describe the calculation of the recommendations.

3.1 DCRS Framework

In OLE, resources combine with learners produce learning activities. Learners' preferences and competencies are extracted from learning activities, plus learners' context information to form learners' models. The constant information of resources are used to form resources profiles. Finally, learners' models and resources profiles are projected into a high-dimensional vector space, the similarities between learners and resources are calculated, and recommendations are made. The overall framework is shown in Fig. 1 below.

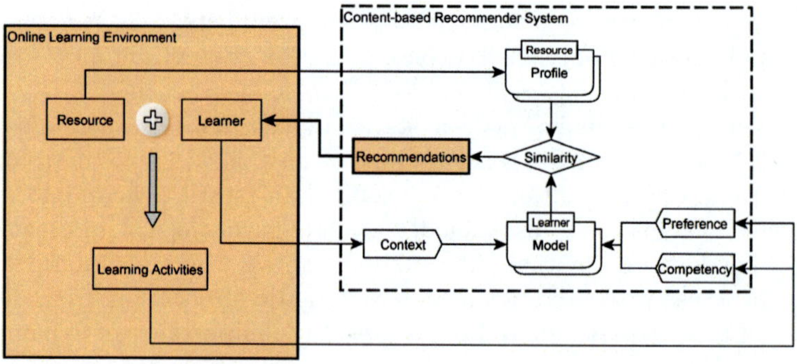

Fig. 1. The overall framework of DCRS

3.2 Activity Records

Users' certain information reflects on their activities, the same in online learning industry, learners select educational resources accordingly to their current preferences and competencies, and whenever learners have new needs, they perform activities accordingly to those needs. Moreover, as long as the learning continues, these learning activities will continue to be produced in real-time.

Learners' dynamic information (preferences and competencies) can be derived from their learning activity records. Educational data mining is an application of data mining to derive meaningful information from educational systems which can be used to improve the learning experience of learners [17]. In this paper, we focus on the activities related to learners' learning, such as, share resources, access resources, download resources, and vote for the resources they used. The vote activity generates a score from 1 to 5, the higher the score, the better the target resource is perceived by the learner. All the activity records will be stored as a data set, and every single record can be explained as "learner ID, knowledge concept, activity, resource name, result", e.g., learner 027, apprentissage profond, vote, *Programmer efficacement en C++*, 5.

3.3 Dynamic Learner Model

This part mainly talks about the three dynamic factors (preference, competency, and context). Among them, preference and competency information need to be derived from the activity records presented in Sect. 3.2.

Preference. Learners have different learning styles, which is reflected in the fact that different learners have different learning preferences. And these preferences can be mainly divided into format type preference (e.g., document-book chapter and document-scientific paper, audio, video, person) and language preference (e.g., English, French, German, Spanish). These preferences can be analyzed and derived from the resources that learners have used in the past. For example, for one target learner, the recent resource usage records of the target learner are selected, the percentage of each preference dimension is calculated, and the final preferences of that target learner will be presented as distributed preference dimensions with percentages.

Competency. First, different learners have different degrees and different learning domains, and these two dimensions can reflect the basic knowledge levels of learners. In addition, the resources shared by the target learner, can be scored by other learners with different degrees (professor and associate professor, PhD student, master student, engineer and undergraduate student), which is also a important method to evaluate the target learner's competency. It is also important to mention that the learners themselves can be recommended to other learners as a special learning resource, the more times a learner is recommended as a resource, the more competent that learner is. So, this is also a dimension for competency evaluation.

Finally, we propose the competency evaluation rules, shown as Table 1, the left side of the table are the defined influence factors, the right side are the corresponding weights.

Table 1. The influence factors and weights of competency evaluation

Influence factors	Weight
Learner's own degree	w_p
Learner's own learning domain	w_{ld}
Shared resources scored by (associate) professor	w_{sbp}
Shared resources scored by PhD student	w_{sbd}
Shared resources scored by mater student	w_{sbm}
Shared resources scored by engineer or undergraduate student	w_{sbe}
Learner recommended to others as a resource	w_{rar}

Suppose that the resources uploaded by target learner i about knowledge concept j are used and scored by α (associate) professors, β PhD students, γ master students, and δ engineer (undergraduate) students. The number of times target learner i is recommended as a resource to other learners is ϵ. The competency, $c_{l_i}^{k_j}$, of learner i, l_i, about knowledge concept j, k_j, is calculated by Eq. 1:

$$c_{l_i}^{k_j} = w_p + w_{ld} + \sum_{n=1}^{\alpha} v_n * w_{sbp} + \sum_{n=1}^{\beta} v_n * w_{sbd} + \sum_{n=1}^{\gamma} v_n * w_{sbm} + \sum_{n=1}^{\delta} v_n * w_{sbe} + \epsilon * w_{rar} \quad (1)$$

where v_n means the scores voted by corresponding learners, obtained from the records presented in Sect. 3.2. $w_p - w_{rar}$ are the weights in table 1.

Context. Learners will have different levels of concentration in different contexts, and different types of resources can stimulate learners' concentration to different levels [5]. It is reasonable to recommend different types of resources to respond to changes in contexts, so as to maintain learners' concentration at high levels. The connections between context, concentration, and resource types are shown in Fig. 2. For the current stage of the research, we only consider the three most significant context factors, location, time, and temperature, and we will consider more factors in the future research.

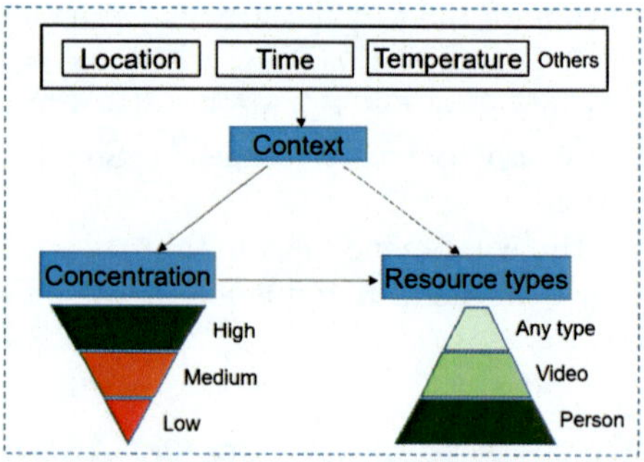

Fig. 2. The connections between context, concentration, and resource types

The recommendation rules for resource types are shown as Table 2 below, where A means the location of learner is quite (e.g., dB $\leq$ 70), B means the time is suitable (e.g., 09:00-11:00, 14:00-16:00), C means the temperature is comfortable (e.g., 18 °C-25 °C). + means that the conditions are met at the same time, / means that only one side of the conditions is met, 0 means no conditions are met. When all three conditions of A, B, and C are met, it means that the concentration is high, and the learner is suitable for any type of resources; when two of the conditions are met, it is inferred that the concentration is at a medium level, and the learner is suitable for more stimulating resources, such as video or audio courses; when only one condition is met or all conditions are not met, it means that the concentration is low, and it is recommended that learners learn in the form of real-time interaction with people, since we presented in paragraph.

Competency. That learners can be recommended to other learners as special resources.

Table 2. The recommendation rules for resource types

Contexts	Concentration resource-types
$A + B + C$	High any type
$A + B/A + C/B + C$	Medium video/audio
$A/B/C/0$	Low person

Dynamic Learner Model. Dynamic learner model contains three components, context, preference, and competency. For the competency part, learners have a corresponding competency value about each knowledge concept, due to learners can study resources on multiple knowledge concepts at the same time. The structure of dynamic learner model is shown as Fig. 3.

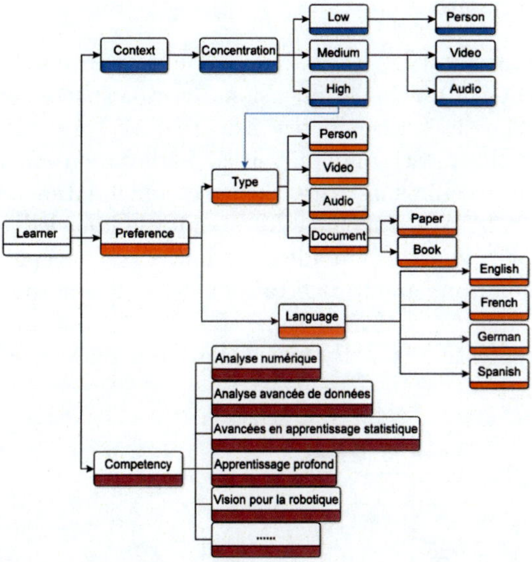

Fig. 3. The structure of dynamic learner model

3.4 Resource Profile

The resource profile also contains three components and ten dimensions, defined as Fig. 4 below, corresponding to the learner model, and the values of all dimensions are determined by the sharer (the learner who share this resource). Since learners intend to share certain resources, then they must know the resources well, otherwise, if the learners shared the resources without fully understanding them, these sharing activities are irresponsible.

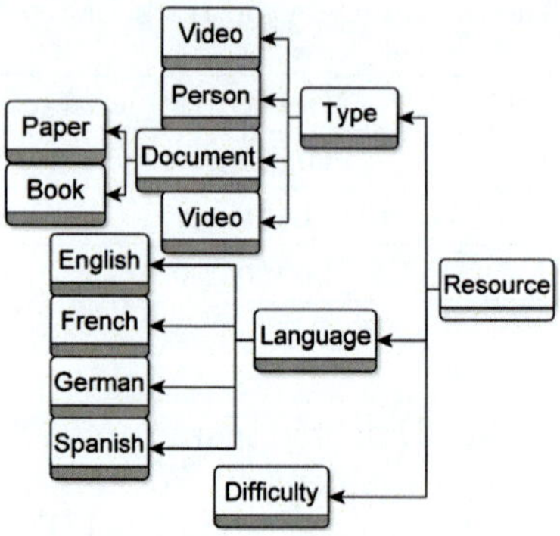

Fig. 4. The structure of resource profile

3.5 Recommendation

The calculation process of DCRS is shown as Fig. 5. Projecting learner models and resource profiles into a high-dimensional vector space, getting their vector representations. The similarities between learner vectors and resource vectors are calculated by the most commonly used high-dimensional vector distance calculation method, cosine similarity, then recommendations are made according to the similarities. A learner may have multiple knowledge concepts, and a resource may be also shared for multiple knowledge concepts, therefore, when making recommendations about a certain knowledge concept, learner's competency must correspond to that knowledge concept. And for another hand, when the learner's concentration is high, the calculation is performed according to the

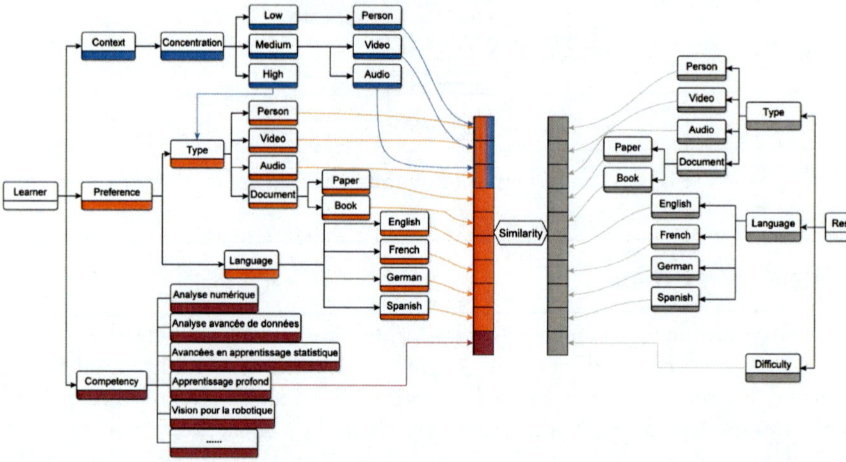

Fig. 5. The calculation process of DCRS

learner's original preferences, but when the learner's concentration is medium or low, the calculation is performed according to the rules presented in Table 2.

4 Discussion:Dynamic Factors Integration in Content-Based Recommender System (CB RS)

Integrating dynamic factors within conventional CB RSs can help conquer the challenges faced by conventional CB RSs, which has obtained significant achievements in OLEs [18]. The CB RSs work with the description of resources and learners, the recommendations of resources are based on what learners already liked. However, the reality is learners' preferences are highly subjective, will change as their emotions shift; learners' competencies are variable, a period of non-learning leads to a decrease and a usage of appropriate resource gets a promotion; and learners' learning efficiencies are closely related to concentration, different contexts produce different levels of concentration. So, how to get these dynamic information? It is unreasonable to rely on learners to assist in filling in these information. As mentioned in Sect. 3.2, learners' certain information reflects on their activities, we proposed to obtain preference information (format type, and language) and competency information (for each knowledge concept) from learners' activity records, and to obtain context information (e.g., location-quite?, time-suitable?, temperature-comfortable?) by sensors.

In the defined dynamic learner model, context contains only three factors, further research is needed in the future to explore more factors; the 9 dimensions of preference part are corresponding to the dimensions of type part and language part of resource; and there are several competencies refer to their respective knowledge concepts, resource recommendations on different knowledge concepts, requiring different corresponding competencies. In the defined resource profile, difficulty as the tenth dimension corresponds to the learners' competencies about the same knowledge concept.

5 Conclusion

In this paper, we propose an adaptive online learning resource recommendation framework, i.e., Dynamic Content-based Recommender System (DCRS). It automatically captures learners' changes of preferences, continually calculates learners' competencies by a dynamic mechanism, considers the impact of context changes during the learning process in real time. DCRS incorporates three dynamic factors (preference, competency, and context) that will affect the learners' choices of resources, makes the recommendations more explainable and adaptive.

References

1. Saleh, M., Abel, M.-H.: System of information systems to support learners (a case study at the University of Technology of Compiégne). Behav. Inf. Technol. **36**, 1097–1110 (2018)

2. Chughtai, M.W., Selamat, A., Ghani, I., Jung, J.J.: Retracted: e-learning recommender systems based on goal-based hybrid filtering. Int. J. Distrib. Sensor Netw. **10**(7), 912130 (2014)
3. Zhang, J., Tian, J.: Incorporating social information in recommender systems using hidden Markov model. In: Huang, G., Chien, C.F., Dou, R. (eds.) Proceeding of the 24th International Conference on Industrial Engineering and Engineering Management, pp. 189–196. Springer, Singapore (2019). https://doi.org/10.1007/978-981-13-3402-3_21
4. Koren, Y.: Collaborative filtering with temporal dynamics. Communications **53**(4), 89–97 (2010)
5. Ma, M.Y., Wei, C.C.: A comparative study of children's concentration performance on picture books: age, gender, and media forms. Interact. Learn. Environ. **24**(8), 1922–1937 (2016)
6. Emanuel, E.J.: Online education: MOOCs taken by educated few. Nature **503**(7476), 342–342 (2013)
7. Zhang, W., Qin, S.: A brief analysis of the key technologies and applications of educational data mining on online learning platform. In: 2018 IEEE 3rd International Conference on Big Data Analysis (ICBDA), pp. 83–86 (2018)
8. Christensen, G., Steinmetz, A., Alcorn, B., Bennett, A., Woods, D., Emanuel, E.: The MOOC phenomenon: who takes massive open online courses and why? (2013)
9. Guo, P.: MOOC and SPOC, which one is better? Eurasia J. Math. Sci. Technol. Educ. **13**(8), 5961–5967 (2017)
10. Aissaoui, O.E., Oughdir, L.: A learning style-based ontology matching to enhance learning resources recommendation. In: 2020 1st International Conference on Innovative Research in Applied Science, Engineering and Technology (IRASET), pp. 1–7 (2020)
11. Chen, W., Mizoguchi, R.: Communication content ontology for learner model agent in multi-agent architecture. In: Proceedings of AIED99 Workshop on Ontologies for Intelligent Educational Systems (1999)
12. Vagale, V., Niedrite, L.: Learner model's utilization in the e-learning environments. In: DB&Local Proceedings, pp. 162–174 (2012)
13. Lu, J., Wu, D., Mao, M., Wang, W., Zhang, G.: Recommender system application developments: a survey. Decis. Support Syst. **74**, 12–32 (2015)
14. Aguilar, J., Valdiviezo-Díaz, P., Riofrio, G.: A general framework for intelligent recommender systems. Appl. Comput. Inform. **13**(2), 147–160 (2017)
15. Chen, W., Niu, Z., Zhao, X., Li, Y.: A hybrid recommendation algorithm adapted in e-learning environments. World Wide Web **17**(2), 271–284 (2012). https://doi.org/10.1007/s11280-012-0187-z
16. Takano, K., Li, K.F.: An adaptive e-learning recommender based on user's web-browsing behavior. In: IEEE 2010 International Conference on P2P, Parallel, Grid, Cloud and Internet Computing, pp. 123–131 (2010)
17. Desmarais, M.C., d Baker, R.S.: A review of recent advances in learner and skill modeling in intelligent learning environments. User Model. User Adap. Interact. **22**, 9–38 (2012)
18. Kardan, A.A., Ebrahimi, M.: A novel approach to hybrid recommendation systems based on association rules mining for content recommendation in asynchronous discussion groups. Inf. Sci. **219**, 93–110 (2013)

Towards Personalized Educational Resources Recommendations for Teachers

Nader N. Nashed[1,2(✉)], Christine Lahoud[1], and Marie-Hélène Abel[2]

[1] Université Française d'Égypte, Cairo, Egypt
{nader.nashat,christine.lahoud}@ufe.edu.eg
[2] Université de Technologie de Compiègne, Sorbonne Universités, Laboratoire Heuristique et Diagnostic des Systèmes Complexes (HEUDIASYC), UMR-CNRS 7253, Compiègne, France
{nader.nashed,marie-helene.abel}@utc.fr

Abstract. Recommender systems experience the challenging and complex process of finding educational resources recommendations for teachers which requires a careful awareness of the teacher's context. These systems should take into consideration the contextual coexistence of teachers in multiple environments or contexts such as social, working, and sentimental contexts. The collected data for each of these contexts is assembled from different sources and systems with different data architectures. Accordingly, this paper provides a multi-step proposition towards a personalized educational resource recommendations for teachers. Furthermore, the paper discusses the data sources from which the contextual data is collected, in addition to the purposed methodology for integrating and mapping the data into the ontology. At the end, the paper illustrates the importance and the role of the teacher analytics and the sentimental data in this proposition. The proposed approach results in a complete representation of the teacher's context and hence, the personalized recommendations.

Keywords: Recommender systems · Education · Ontology · Teacher

1 Introduction

Recommender systems (RSs) filter information to predict user's preferences towards a specific item. Through this process, it imitates how the social exchange of recommendations works. During the 90s and 00s, RSs were developed mainly for commercial applications [15]. In an educational context, RSs can provide personalized recommendations of learning resources based on each user's details and interests [25]. Recommendations can be personalized with the help of the complete context description for each user and in our case, teachers. Context description can be defined differently according to the research point-of-view and the predefined actors related to this context. Each research point-of-view

L. Troiano et al. (Eds.): ICDLAIR 2021, LNNS 441, pp. 107–116, 2022.
https://doi.org/10.1007/978-3-030-98531-8_11

tends to provide its own unique individualistic perspective which can be completely different from other research fields concerning the same context actors. Also in the educational field, context can be defined from different point-of-views: geography, educational institution structure, learner, or teacher. Since this research focuses on educational resources recommendations for teachers, we limit our scope to teacher's context in terms of profile, demography, level of teaching, experiences, social environment, cultural background, and working environment [16].

In this paper, we purpose an approach to provide personalized educational resources recommendations for teacher. The proposed approach introduces a context-aware methodology for the recommendations using an ontology-based educational data integration [20]. The approach covers multiple contexts of a teacher, such as living, working [19], and sentimental contexts [21]. The remaining of this paper is organized as follows: Sect. 2 states the current social and scientific challenges in our research. Section 3 explores the related work to this paper. Section 4 introduces the proposed approach and its architecture. Section 5 discusses the usability and the usefulness of this approach with illustrated examples. Finally, Sect. 6 summarizes and concludes this paper and provides its perspectives.

2 Problem Statement

A context-aware recommendation can be achieved through a proper definition of the existing problem. At the beginning, teacher contexts must be selected according to the relation with the teaching/learning performance. Between the social representation and the multiple identities of a teacher, we can summarize the teacher's coexistence into the following identities: a living/cultural identity and a working/institutional identity. Identities are a way of providing an explanation of meanings and experiences of a certain person based on cultural attributes that are interrelated with other forms of meanings [6]. These meanings are inseparable when it comes to the associated context with a certain identity, such as a teacher's identity definition. Other contexts are said to be in a continuous process of reconstruction due to the dynamicity of this entity with the surrounding conditions. One of these contexts is the sentimental context of a person and more preciously, the teacher's sentimental wellbeing [17]. Therefore, we need to identify the effective contexts for a teacher taking into consideration living and working environments in addition to the sentimental state.

A two-pronged solution is need in this case by well-represented of the multiple contexts of a teacher in addition to the data integration and processing according to these contexts. A teacher coexists in multiple contexts in real-life such as living context, working context and sentimental context. The solution must consider the well-representation of all contexts as well as the integration process between them. For well-representation, each of the three contexts must be defined at the beginning. The living context is described as the birth environment in which the teacher was born and raised and the living environment where he lives during the current moment. The working context is represented as the working conditions in the educational institution where the teacher works in addition to the living conditions of the area demographically and geologically. The sentimental context can be defined as the emotional status of the teacher during the daily life which is affected by multiple stimuli [22]. Mood can perfectly describe the explainable and unexplainable affective states simultaneously [11]. In order to provide context-aware recommendation, all the previously mentioned contexts should be carefully described and represented for smooth data treatment.

Ontology representation solves the first scientific challenge as it provides an enhanced visualization of data and relationships in addition to the context sharing between multiple information systems. Another problem arises during the integration of the contexts because of the different formats and schemas of the collected data from each context. For an optimal integration of data, we need to discover all the available data concerning our cause and then, to identify the relevant ones. On the other hand, the discovered data will be meaningless if it was not analyzed, and insights were extracted. Therefore, it can be said that data integration challenge results into two sub-challenges: discovery and exploitation [13]. A probable solution is the ontology-based data integration providing uniqueness of data and facilitating the data insertion and retrieval. Integrating these two approaches can present a sufficient solution for the discussed problems.

3 Related Work

The related research to this paper, is divided into two categories of research to address the two parts of the proposed two-pronged solution. The first category is the educational data integration and the data modeling unification targeting the educational resources. The second category is the context-related educational ontologies and its related educational resources recommender systems.

3.1 Educational Data Integration

The educational data integration approaches have multiplied during the past decade targeting the different data categories. The review conducted by Desmarais et al., explores the data modeling and integration approaches from the learner's point-of-view [8]. Through this review, it was noticed that none of the reviewed models was targeting the teacher's point-of-view. Dhekne et al. introduced a new Massive Open Online Courses (MOOC) aggregation from different providers [9]. The similar approach was followed by Ambite et al. to link and index training resources using a unified data model [3]. The latest research that was done by Duran et al. [10], integrates open educational resources (OERs) from various sources that are related to Computing and Electronics engineering field using RDF and ontology representation. We can observe the lack of educational data integration research from the teacher point-of-view.

3.2 Educational Ontologies and Ontology-Based Recommender Systems

The research on the context representation in the educational domain had common interests in the learner point-of-view or the administrative structure of the educational institution. The systematic review conducted by Costa et al. [7], shows the significant amount of research directed towards the educational context in terms of learning analytics and context description. The teacher context is almost absent from the reviewed papers. For example, Guettat et al. introduced a computational domain ontology to profile the student/learner and the corresponding relation with the learning resources only without any mention of the teacher's interconnecting relationship in this context [14]. While the research conducted by Antonelli et al., implements an ontology to represent an open network platform for higher education [5]. The teacher was placed in this ontology from the learning outcomes perspective only and the authors illustrated the relation between the expected learning outcomes, the learning process, and the learners without further linkage to any contextual concepts.

The state-of-the-art article by Stancin et al. [23], reviewed more than 90 ontologies in the educational context that were published between 2015 and 2019, none of these ontologies was constructed from the teacher's point-of-view. Ibrahim et al. implemented three educational ontologies: course, student, and job ontologies [18]. The teacher and the contextual representation were absent from the three introduced ontologies. Another remarkable ontology from Stancin review, is the ontology created by Fuchs et al. [12]. This ontology represents learners, learning content, and learning semantics and it does not represent neither the learners' context nor the relationship with their teacher/instructor.

The comprehensive review performed by Tarus et al. [24], collects the ontology-based recommender systems in the educational domain from 2005 until 2014. Most of the reviewed journal papers targeted learners or the educational system in general and the used ontologies were from the learner or the system point-of-view. We conclude that the research in the educational domain lacks concentration on the teacher point-of-view. Therefore, research is needed in the context description ontologies for teacher as well as the ontology-based data integration in the same domain.

4 Approach Architecture

The proposed recommender system model consists of multiple steps as shown in Fig. 1. During the first step, the data is collected and integrated from multiple relational data sources. These data sources have various formats with different structure levels from multiple origins. When data is collected and integrated from different sources, it reflects the dispersed contexts of the teacher in a more precise description and observations. Data integration techniques, such as conflict resolution function and cosine similarity, prevents data redundancy and conflictions. Four datasets were integrated into one data collection to represent the living and working contexts of the teacher. As for the learning resources, another six datasets were integrated from different data sources and with different formats.

Through the second step, the relationally integrated data is mapped to the ontology. Ontology is a formal representation of a conceptualization of a certain domain that provides explicit domain assumptions. Also, the ontology facilitates the knowledge sharing among multiple sub-systems, as well as the reusability, communication, and collaboration. The proposed ontology is constructed based on the teacher's multiple contexts and his preferences; teacher profile, learner profile, living environment and working environment. The third and fourth steps concern with the hybrid recommender engine, and the filtering and sorting of the resulting recommendations. The fifth step of this proposal deals with the teacher behaviors and interactions with the system. The direct interactions with resources, concepts, and knowledge sharing are recorded for each teacher using the MEMORAe SoIS platform [1]. However, the indirect interaction through the teacher's sentimental state (mood), are captured using the Moodflow@doubleYou platform [4]. The mood is detected according to the 3-level classification approach: positive, negative, and neutral. The teacher's interactions and mood act as the user feedback to resort and filter the provided list of recommendations from the fourth step. The emotion-based data introduces a new dimension of personalization and draws the system's attention to new insights of teacher's preferences. Therefore, the recommender system can provide the teacher with enjoyable resources, and it can evaluate the recommendations correctly.

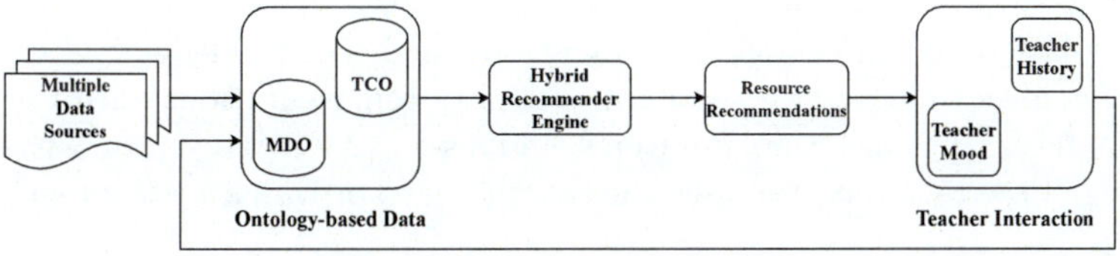

Fig. 1. Block diagram of the model architecture

4.1 Ontology-Based Data

Teachers coexist in multiple contexts which leads to a data extraction process from multiple sources with different architecture and formats. These different sources can be divided into four main categories: educational resources, personal information, environmental data, and educational institutional data as shown in Fig. 2. All the data categories were integrated and mapped into the different ontologies [20]. Only one category was not covered in the mentioned work, the personal data, which is collected during real-time interaction of a teacher with an educational recommender system.

Afterwards, the integrated data is mapped into the integrated ontologies: Teacher-Context Ontology (TCO), MEMORAe ontology, and Mood Detection

Ontology (MDO) [19,21]. The three ontologies extend one another to complete the full representation of a teacher's context as shown in Fig. 3. TCO represents teacher coexistence in three different contexts: work environment, living environment, and educational institution, while MDO provides a sentimental state representation within the teacher-resource interactivity. These interactivities, which are part of the integration between TCO and MEMORAe (MCC) ontologies, affects the teacher's mood and therefore, his working performance. On the other hand, MCC ontology represents the user account and knowledge sharing within an organization which enforces the activities tracking and placement within the teacher's context. These integrated contextual concepts are used by the hybrid recommender engine to provide personalized educational resources' recommendation to the targeted teacher.

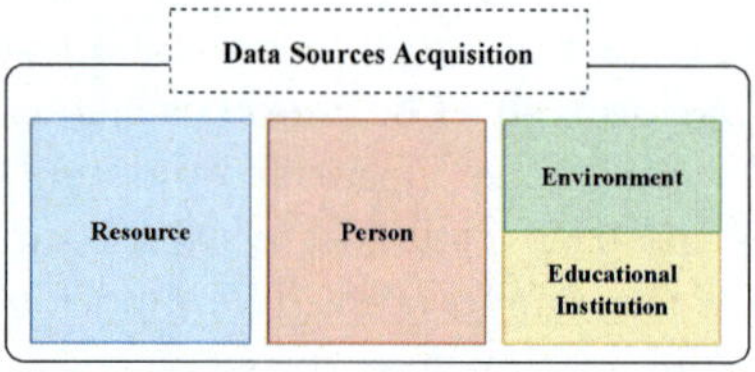

Fig. 2. Data categories in teacher context

4.2 Hybrid Recommender Engine

Ontology-based recommender engines are defined as knowledge-based recommender engines that replace the relational data representations with the ontology semantic representation for items and users [2]. In this proposition, the knowledge-based engine processes the search query provided by the teacher and

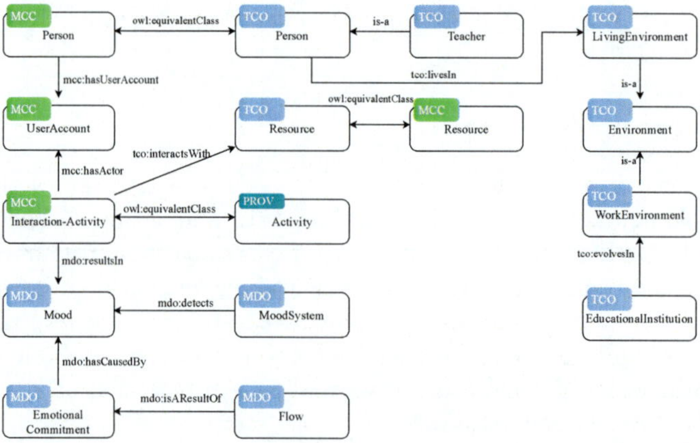

Fig. 3. TBox ontology representation of the multiple teacher contexts

evaluates the resulting list according to the teacher's profile and previous activities. The cosine similarity function calculates the relative score for each feature-search parameter couple and afterwards as in Eq. 1 where R_a is a search list and R_b is a previously rated resource for all n resource descriptive features, the Pearson utility function evaluates each feature's importance to this specific teacher [2]. The resulting list of recommendation is forwarded to the filtering and sorting algorithm to reevaluate the recommendations with respect to the context-aware recommender engine.

$$Sim(R_a, R_b) = \frac{\sum_{i=1}^{n}(R_{ai}R_{bi})}{\sqrt{\sum_{i=1}^{n} {R_{ai}}^2}\sqrt{\sum_{i=1}^{n} {R_{bi}}^2}} \tag{1}$$

The context-aware recommender engine is responsible for matching a teacher's contextual situation with another teacher. But at the beginning, the engine extracts the contextual dimensions/factors for a certain teacher and initiates the teacher-teacher matching process. These dimensions/factors are dependent on the person's situation which prevents the unification of the selected factors. Moreover, predictions of context-aware recommender systems, can be imprecise if all contextual factors were weighted equally. The personalized process of weighting and selection of contextual factors improves the recommendations' quality according to each teacher. Therefore, the proposed approach

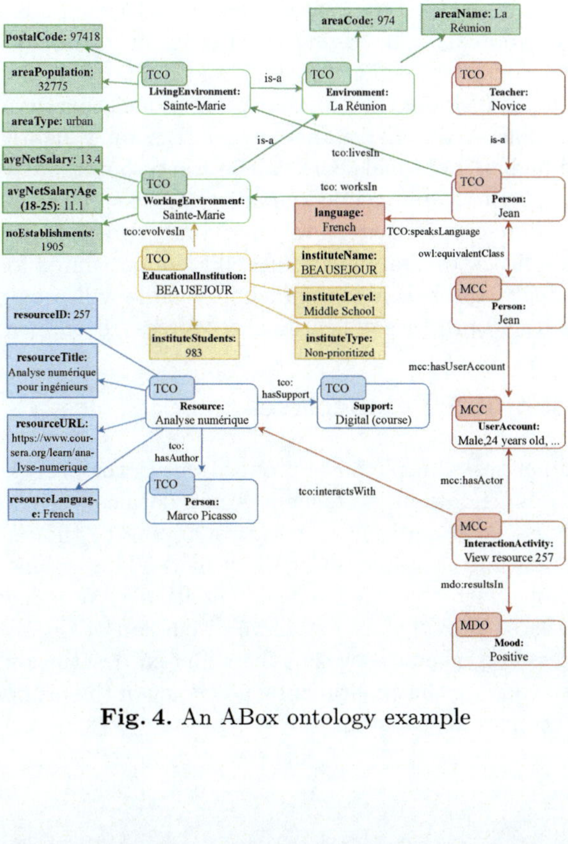

Fig. 4. An ABox ontology example

introduces a methodology to weight and select contextual factors relevant to a certain teacher's context.

The Semantic Web Rule Language (SWRL) is a rule markup language that is used with Ontology Web Language (OWL) to express logical rules into readable syntax. In this approach, SWRL rules are used to sort the teachers' peer list generated from both recommender engines. For each teacher, the rules are applied to reason the ABox instance corresponding to him. A total score is calculated by counting the number of satisfied rules. Then, the list of resources' recommendations is sorted according to a descending order of the obtained scores. At the end, the teacher is asked to view and interact the recommended resources. The mood of the teacher is detected during the interaction process and is used to evaluate further recommendations process.

5 Discussion

The ABox representation of the novice teacher, Jean, is illustrated in Fig. 4. The teacher contexts representation along with the sentimental feedback empower both recommender engines. The knowledge-based engine uses the teacher's profile information and searches for resources/items match. In our example, Jean, who speaks French and obtained an engineering degree, is searching for a course about "Numeric analysis", therefore, one of the recommended resources is the French course, "Analyse numérique". On the other hand, the context-aware engine extracted and weighted the context factors and then, finds another teacher similar to Jean's context. The engine finds that most similar teachers speak French, and they coexist in similar working and living environments. Also, the context-aware engine includes the "Analyse numé'rique" course within the resulting recommendations list. Finally, the sorting reasoning algorithm scores this course with the highest obtained score. When the resource was viewed by Jean, he was satisfied by the recommended course which is proved by the generated positive mood. Therefore, the extensive contextual representation of the teacher opens new possibilities to grant a full-usability opportunity for recommender engines. Accordingly, the hybrid recommender engine will provide each teacher with personalized and precious educational resources' recommendations.

6 Conclusion and Perspectives

This paper summarizes multiple novel propositions in the process of introducing personalized recommendations to teachers. A new data collection and integration approach is discussed, in addition to a data mapping to the ontology. Teacher analytics collection and sentimental analysis methods are illustrated to monitor the interactions with the given recommendations. We believe that none of the previous research provided an extensive discussion of the recommendation-generation process for teachers as well as their context description. For the future work, we need to validate the experimental efficiency of the proposed approaches with a hybrid recommender system.

References

1. Abel, M.-H., Benayache, A., Lenne, D., Moulin, C.: E-MEMORAe: a content-oriented environment for E-learning. In: Pierre, S. (ed.) E-Learning Networked Environments and Architectures. AIKP, pp. 186–205. Springer, London (2007). https://doi.org/10.1007/978-1-84628-758-9_7
2. Adomavicius, G., Tuzhilin, A.: Toward the next generation of recommender systems: a survey of the state-of-the-art and possible extensions. IEEE Trans. Knowl. Data Eng. **17**(6), 734–749 (2005)
3. Ambite, J.L., Gordon, J., Fierro, L., Burns, G., Mathew, J.: Linking educational resources on data science. In: Proceedings of the AAAI Conference on Artificial Intelligence, vol. 33, pp. 9404–9409 (2019)
4. Andres, F., Blancan, B., Foutouhi, F., Ghinea, G., Leite, M.C.A.: Overview of the 4th DECOR workshop 2021. In: 37th IEEE International Conference on Data Engineering (2021)
5. Antonelli, D., et al.: Tiphys: an open networked platform for higher education on industry 4.0. Proc. CIRP **79**, 706–711 (2019)
6. Castells, M.: The Power of Identity, vol. 14. Wiley, Chichester (2011)
7. Costa, L.A., Pereira Sanches, L.M., Rocha Amorim, R.J., do Nascimento Salvador, L., dos Santos Souza, M.V.: Monitoring academic performance based on learning analytics and ontology: a systematic review. Inform. Educ. **19**(3), 361–397 (2020)
8. Desmarais, M.C., d Baker, R.S.J.: A review of recent advances in learner and skill modeling in intelligent learning environments. User Model. User-Adap. Interact. **22**(1), 9–38 (2012). https://doi.org/10.1007/s11257-011-9106-8
9. Dhekne, C., Bansal, S.K.: MOOClink: an aggregator for MOOC offerings from various providers. J. Eng. Educ. Transform. (2018)
10. Durán, C.G., Ramírez, C.M.: Integration of open educational resources using semantic platform. IEEE Access **9**, 93079–93088 (2021)
11. Frijda, N.H., Scherer, K.R.: Emotion definitions (psychological perspectives). In: The Oxford Companion to Emotion and the Affective Sciences, pp. 142–144 (2009)
12. Fuchs, K., Henning, P.A.: Cognitive space time: a model for human-centered adaptivity in E-learning. In: 2018 IEEE International Conference on Engineering, Technology and Innovation (ICE/ITMC), pp. 1–9. IEEE (2018)
13. Golshan, B., Halevy, A., Mihaila, G., Tan, W.-C.: Data integration: after the teenage years. In: Proceedings of the 36th ACM SIGMOD-SIGACT-SIGAI Symposium on Principles of Database Systems, pp. 101–106 (2017)
14. Guettat, B., Farhat, R.: An approach to assist learners to identify their learning objectives in personal learning environment (PLE). In: 2015 5th International Conference on Information & Communication Technology and Accessibility (ICTA), pp. 1–6. IEEE (2015)
15. Gupta, V., Pandey, S.R.: Recommender systems for digital libraries: a review of concepts and concerns. Libr. Philos. Pract. 1–9 (2019)
16. Han, J., Yin, H.: Teacher motivation: definition, research development and implications for teachers. Cogent Educ. **3**(1), 1217819 (2016)
17. Heutte, J., Fenouillet, F., Martin-Krumm, C., Boniwell, I., Csikszentmihalyi, M.: Proposal for a conceptual evolution of the flow in education (EduFlow) model. In: 8th European Conference on Positive Psychology (ECPP 2016) (2016)
18. Ibrahim, M.E., Yang, Y., Ndzi, D.L., Yang, G., Al-Maliki, M.: Ontology-based personalized course recommendation framework. IEEE Access **7**, 5180–5199 (2018)

19. Nashed, N.N., Lahoud, C., Abel, M.-H.: TCO: a teacher context ontology. In: 2021 IEEE 24th International Conference on Computer Supported Cooperative Work in Design (CSCWD), pp. 757–762. IEEE (2021)
20. Nashed, N.N., Lahoud, C., Abel, M.-H.: Ontology-based teacher-context data integration. In: 2022 IEEE/SICE International Symposium on System Integration (SII 2022). IEEE/SICE (2022)
21. Nashed, N.N., Lahoud, C., Abel, M.-H., Andrès, F., Blancan, B.: Mood detection ontology integration with teacher context. In: 2021 IEEE 20th International Conference on Machine Learning and Applications. IEEE (2021)
22. Russell, J.A., Barrett, L.F.: Core affect, prototypical emotional episodes, and other things called emotion: dissecting the elephant. J. Pers. Soc. Psychol. **76**(5), 805 (1999)
23. Stancin, K., Poscic, P., Jaksic, D.: Ontologies in education-state of the art. Educ. Inf. Technol. **25**, 5301–5320 (2020). https://doi.org/10.1007/s10639-020-10226-z
24. Tarus, J.K., Niu, Z., Mustafa, G.: Knowledge-based recommendation: a review of ontology-based recommender systems for e-learning. Artif. Intell. Rev. **50**(1), 21–48 (2017). https://doi.org/10.1007/s10462-017-9539-5
25. Xiao, J., Wang, M., Jiang, B., Li, J.: A personalized recommendation system with combinational algorithm for online learning. J. Ambient. Intell. Humaniz. Comput. **9**, 667–677 (2018). https://doi.org/10.1007/s12652-017-0466-8

DEEC and EDEEC Routing Protocols for Heterogeneous Wireless Sensor Networks: A Brief Comparative Study

Kenza Redjimi, Mehdi Boulaiche, and Mohammed Redjimi(✉)

Department of Computer Science, LICUS Laboratory, Université 20 Août 1955, 21000 Skikda, Algeria

{k.redjimi,m.boulaiche,m.redjimi}@univ-skikda.dz

Abstract. Nowadays, humanity lives in the digital era; the machine age. Information is shared between humans and machines ubiquitously. Several researchers focus on data routing in wireless sensor networks (WSN). Among the major challenges to make the network more and more efficient are the conservation of energy, which can extend the life of the network, and the speed of routing of information, which allows the network to respond quickly. The network security is another important aspect in the WSNs. These factors are the fundamental design issues in sensor networks. A WSN protocol can be either homogeneous or heterogeneous. In the homogeneous WSN, the sensor nodes are supplied with an equal amount of initial energy while heterogeneous networks have uneven initial energy distribution. One may consider multi-level initial energy distribution. In a two-level distribution, advanced nodes have higher energy in comparison to normal nodes. Distributed Energy Efficient Clustering (DEEC) and Enhanced Distributed Energy Efficient Clustering (EDEEC) heterogeneous wireless network protocols are presented in this paper.

Keywords: WSNs · Clustering · Heterogeneous protocols · Wireless sensor networks · DEEC · EDEEC

1 Introduction

With the recent advancements in miniaturization techniques and the increased need of services over wireless communications, micro sensors have become a hot research topic [1]. These tiny devices are deployed in a dense manner over a geographic region of interest to form self-configured wireless sensor networks [2]. These networks have a multitude of features that are applied in various fields such as military, health care, security, and in several industries [3, 4]. Each sensor in the network is equipped with sensing units, radio transmission modules, powers suppliers, and limited processing and data storage devices [5]. The principal functionality of these nodes is to gather useful information from the hostile environments and to send them to the far base station (BS) [6]. Many routing protocols have been developed for WSNs. Most of these routing protocols require nodes localization information since a sensed event is directly related

L. Troiano et al. (Eds.): ICDLAIR 2021, LNNS 441, pp. 117–125, 2022.
https://doi.org/10.1007/978-3-030-98531-8_12

to a location. This information is required to calculate distance between two particular nodes such that energy consumption can be estimated. Because there is no addressing system for wireless sensor networks (like IP addressing) and when they are deployed randomly in a region, the location information of these nodes can be used in routing data efficiently in terms of energy. For example, if the region to sense is known by using nodes localization in order to broadcast the request to this particular region, it is easy to eliminate the number of retransmissions.

Among the limitations of WSNs, the sensor nodes have limited initial power energy. Moreover, the batteries cannot be recharged or replaced due to hostile environments where they are deployed. Many research works have been carried out for energy saving of the sensor nodes.

The rest of this article is organized as follows: section two presents some related work on routing protocols in WSNs. Section 3 presents the DEEC and the EDEEC models. Section 4 concerns the simulation results, and Sect. 5 concludes this article.

2 Related Work

A routing protocol for wireless sensor networks consists to find the optimal way and to establish the correct and efficient route between a pairs of nodes by respecting the energy consumption and network lifetime factors [7]. Direct communication between a sensor node and the base station (BS) is not an efficient solution since it consumes a lot of energy and the base station can be located outside the node signal range. Thus, several protocols have been proposed and many of them are implemented. Three important classes of protocols exist: flat, hierarchical, and location-based routing protocols. One can distinguished two categories in flat routing protocols that are On-Demand protocols (such as Dynamic Source Routing (DSR) [8] and Ad hoc On-demand Distance Vector (AODV) [9] or Table-Driven protocols such as Destination Sequenced Distance Vector (DSDV) [10]. In the case of hierarchical protocols, the nodes are grouped together in the form of a cluster and a cluster head is periodically elected to receive the data from the various constituents of the cluster and to aggregate them, and furthermore to route them towards the base station. Low-energy adaptive clustering hierarchy (LEACH) [11] and its variants are among the typical implementations of this class of protocols. An interesting other class of protocols is the location-based routing protocols (so-called geographic protocols). In this case, each node transmits its data to its closets neighbour that is in the direction of the base station. For example, Greedy Perimeter Stateless Routing (GPSR) [12] is a geographic routing protocol. In Geographic Routing, each node is concerned only with its one-hop neighbours.

Sensor nodes in hierarchical architecture generally, are not in direct communication link with the base station. However, the main limitation of WSNs is that the sensor nodes are operating on limited power sources.

One can consider two categories of wireless sensor network's properties homogenous and heterogeneous. In the first category, all the nodes have a same amount of initial energy and the same hardware capability, but in the heterogeneous network, there are different levels of nodes equipped with different amounts of initial energies the hardware requirements are different.

Many protocols for homogeneous and heterogeneous wireless sensor network are developed. Among the first homogenous protocols are LEACH [11], Power-Efficient Gathering in Sensor Information Systems (PEGASIS) [13] and Hybrid Energy-Efficient Distributed Clustering (HEED) [14]. Homogeneous protocol are not efficient in the heterogeneous networks. In heterogeneous WSNs (HWSNs), nodes have different initial energy levels, software and hardware node components may be different or the strength of the node radio signals different. The initial amount of energies of the sensor nodes is the most studied heterogeneity cases. Among this kind of protocols, one can distinguish the following: Distributed Energy Efficient Clustering (DEEC) [15] and its variants: Developed DEEC (DDEEC) [16], Enhanced Developed Distributed Energy- efficient Clustering (EDDEEC) [17] and Enhanced DEEC (EDEEC) [18]. Stable election protocol (SEP) [19], Balanced Energy Efficient Network Integrated Super Heterogeneous (BEENISH) [20] and modified BEENISH [21] are other protocols for heterogeneous WSNs.

3 The Network Model

The following properties are used for simulating sensor region:

- The sensor nodes and the base station are assumed stationary once they are deployed in the environment and their location are fixed.
- Wireless sensor network includes heterogeneous sensor nodes.
- The base station is not limited in terms of energy, memory and computing power.
- The nodes are eligible to determine its current energy level and location information through GPS service.
- All the sensor nodes are immobile and having fixed node id.
- Data aggregation is done only at CH nodes.

The following figure (Fig. 1) shows the initial network vision:

3.1 The Energy Model

The energy consumption model proposed in this paper is depicted in Fig. 2. This model follows the same radio model used in [11]. Therefore, the transmitter dissipates the energy for the transmission of k data bits as the following (Eqs. (1) and (2)):

$$E_{tx}(k, d) = \begin{cases} kE_{elec} + d^2 E_{fs} & \text{if } d < 0 \\ kE_{elec} + d^4 E_{amp} & \text{if } d \leq 0 \end{cases} \tag{1}$$

$$d_0 = \sqrt{\frac{E_{fs}}{E_{amp}}} \tag{2}$$

Where $E_{tx}(k, d)$ is the energy dissipated by the transmitter, E_{amp} is the power amplifier, E_{elec} is the electronic devices energy, d is the transmission distance, E_{fs} is the amplification energy in the free space, d_0 is the threshold distance that depends on the environment and k represents the number of the transmitted bits.

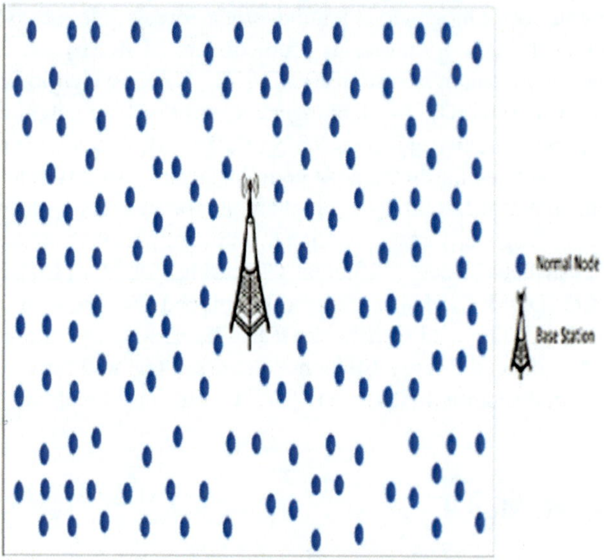

Fig. 1. The network model

Equation (3) gives the energy dissipated by the receiver for k-bits of data:

$$E_{rx} = kE_{elec} \tag{3}$$

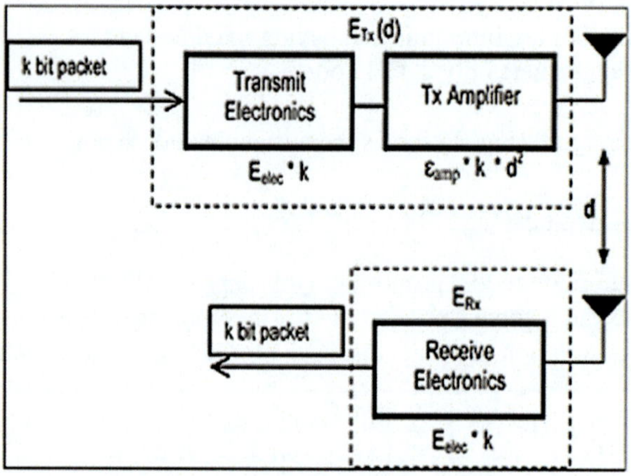

Fig. 2. The used radio energy model.

In a hierarchical architecture, sensor nodes are, generally not, in direct communication link with the base station. At the level of each cluster, a cluster head (CH) is periodically elected (at each round) and all the other nodes of the cluster send their information to it. Thus, among the roles of the CH, is to take care of the aggregation of the data received and their transmission to the base station (BS).

3.2 The Heterogeneous Network Model

Heterogeneous routing protocols for WSNs are categorized by several metrics such as energy efficiency, number of heterogeneity level, stability, and cluster head (CH) selection. Effective energy management is used for choosing the proper cluster head effectively, such as it upgrades the network lifetime up to a certain extent.

The Distributed Energy Efficient Clustering Protocol for Heterogeneous Wireless Sensor Networks (DEEC)

Li Qing, Qingxin Zhu and Mingwen Wang [15] proposed the Distributed Energy Efficient Clustering Protocol for Heterogeneous Wireless Sensor Networks (DEEC), which is a two-level clustering algorithm for heterogeneous WSN. The DEEC routing protocol is applied for heterogeneous WSNs. DEEC uses two energy levels: normal for nodes with low energy and advanced for nodes equipped with higher energy levels. To select a CH, DEEC uses initial and residual energy levels of nodes.

The cluster heads are selected by a probability based on the ratio between residual energy of each node and the average energy of the network. There are different amount of initial energies for the nodes of the sensor network.

Let E_0 and $(1 + a)E_0$ be the initial energies of a normal and an advanced nodes (*a* denotes how many times energies of advanced node has been relative to normal node). Let N be the total number of sensors in the network (4).

$$N = N_{nml} + N_{advc} = N(1 - m) + Nm \tag{4}$$

Where N_{nml} and N_{advc} represent the total number of normal and advanced nodes respectively and m the fraction of the advanced nodes $(0 < m < 1)$.

The total first energies of the normal and advanced nodes are given respectively by Eq. (5), (6):

$$E_{nml} = N_{nml}E_0 \tag{5}$$

$$E_{advc} = N_{advc}E_0 + aN_{advc}E_0 \tag{6}$$

Where a is a multiplicative factor of the initial energy E_0 and a > 1.

Equation (7) calculates the total first energy:

$$\begin{aligned} E_{total} = E_{nml} + E_{advc} &= N(1 - m)E_0 + Nm(E_0 + aE_0) = NE_0(1 - m + m + ma) \\ &= NE_0(1 + ma) \end{aligned} \tag{7}$$

Equation (8) gives the average energy of the network for the rth round:

$$E_{avg} = \frac{1}{N}E_{total}\left(1 - \frac{r}{R}\right) \tag{8}$$

E_{total} is the total energy of all the nodes (the N nodes), R is the number of rounds for the alive network. E_{round} represents the energy consumed for each round (Eq. 9).

$$R = \frac{E_{total}}{E_{round}} \tag{9}$$

Let P_i be the probability of a node S_i to become CH in a round r, G the set of eligible nodes at round r. S_i belongs to G if it becomes CH in recent rounds. At the beginning of each round, each node proposes a random probabilistic number (between 0 and 1) and if this number is less than the threshold $T(S_i)$ then S is eligible to become a CH (Eq. 13).

For the DEEC protocol, the probability for normal and advanced node is given by the following Eq. (12). P_{opt} is a determinate value that represents the fraction of the cluster numbers in the network ($0 < P_{opt} < 1$). p_{opt} can be calculated by using (10) and (11), where N and M are the number of the network nodes and M the size of this this network (M × M) respectively.

$$k_{opt} = \sqrt{\frac{N}{2\pi}} d_0 \frac{M}{d^2} \tag{10}$$

$$p_{opt} = \frac{k_{opt}}{N} \tag{11}$$

$$P_i = \begin{cases} \frac{E_i(r)P_{opt}}{(1+am)E_{avg}} & if\ normal\ node \\ \frac{E_i(r)P_{opt}(1+a)}{(1+am)E_{avg}} & if\ normal\ node \end{cases} \tag{12}$$

$$T(S_i) = \begin{cases} \frac{P_i}{1-P_i(\bmod (r, \frac{1}{P_i})} & if\ S_i \in G \\ 0 & otherwise \end{cases} \tag{13}$$

The Enhanced Distributed Energy Efficient Clustering Protocol for Heterogeneous Wireless Sensor Networks (EDEEC)

Parul Saini, Ajay. K. Sharma [18] proposed the Enhanced Distributed Energy Efficient Clustering protocol (EDEEC). This protocol is three level clustering algorithm to increase the lifetime and stability of the network. The principle of functioning of this protocol is similar to DEEC. However, there are three kinds of nodes: normal, advanced and super nodes (Eqs. (14–17):

$$E_{nml} = NE_0(1 - m) \tag{14}$$

$$E_{advc} = NmE_0(1 - m_0)(1 + a) \tag{15}$$

$$E_{sup} = Nmm_0E_0(1 + b) \tag{16}$$

$$E_{total} = NE_0(1 + m(a + m_0b)) \tag{17}$$

Where a, b are a multiplicative factors of the initial energy E_0 and a, b and b > a, m and‘ m_0 fractions of super and advanced nodes in the network respectively.

The probability P_i is calculated according to Eq. (18):

$$P_i = \begin{cases} \frac{E_i(r)P_{opt}}{(1+m(a+m_0b))E_{avg}} & \text{if normal node} \\ \frac{E_i(r)P_{opt}(1+a)}{(1+m(a+m_0b))E_{avg}} & \text{if advanced node} \\ \frac{E_i(r)P_{opt}(1+b)}{(1+m(a+m_0b))E_{avg}} & \text{if super node} \end{cases} \quad (18)$$

E_{avg} is calculated according to Eq. (8)

The threshold $T(S_i)$ for S_i to be eligible to become a CH is the same as in Eq. (11).

4 Simulation

For the system simulation, the radio parameter values mentioned in Table 1 are used in the DEEC and EDEEC protocols. Advanced nodes have a = 1.5 times more energy and supernodes have b = 3 times energy than normal nodes. In this paper, we are interested essentially with the network lifetime and the packet transmission. Figures 3 and 4 clearly show that the EDEEC protocol is significantly more efficient than DEEC in terms of network lifetime and packets transmitted to the BS.

Table 1. Simulation parameters

Parameters	Values
Network size	(100, 100) m^2
Number of nodes	100
Normal nodes initial energy (E_0)	0.5 J
A	1.5
B	3
Eda (data aggregation)	5 nj/bit
E_{elec}	50 nj/bit
E_{fs}	10 nj/bit/m^2
E_{amp}	0.0013 pj/bit/m^4
d_0	76 m
P_{opt}	0.1
Sink position	(50, 50) in the centre of the network
Fraction of super nodes (m)	0.3
Fraction of advanced nodes (m_0)	0.3

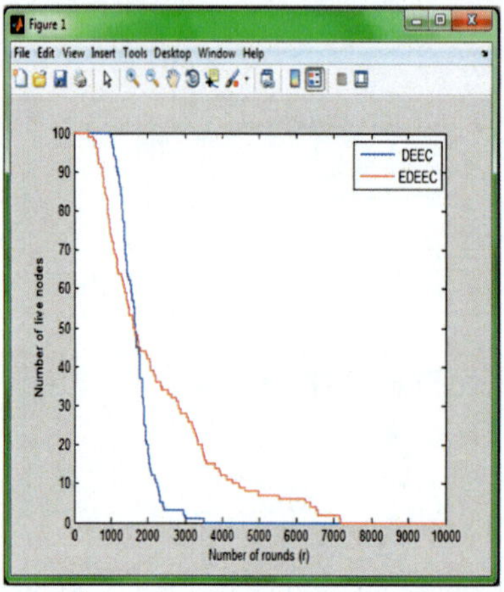

Fig. 3. Number of alive nodes.

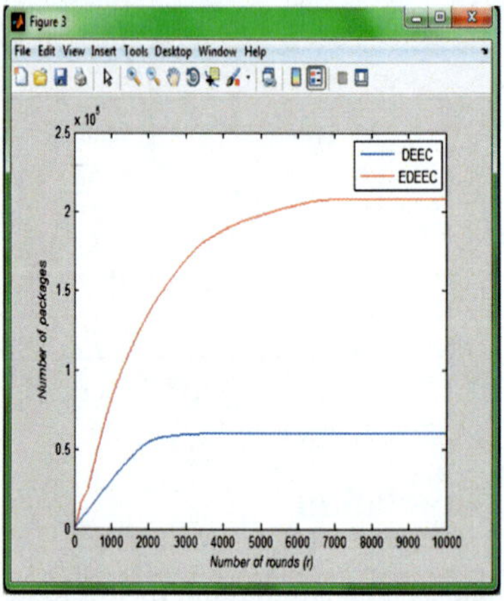

Fig. 4. Number of transmitted packets

Simulation results show that the EDEEC protocol is more efficient than DEEC in terms of the network lifetime and the number of the transmitted packets to the BS. The EDEEC protocol increases the network lifetime of the network. In this simulation, the lifetime of the network and the number of delivered packets to the Base Station are multiplied by 2.

5 Conclusion

In this paper, the DEEC and EDEEC heterogeneous WSN protocols are presented. The simulation results show that EDEEC is significantly more efficient than DEEC. This study was only interested in the case of heterogeneity at two and three levels. In the next work, we propose to address other protocols with higher levels of heterogeneity.

References

1. Zhu, J., et al.: Development trends and perspectives of future sensors and MEMS/NEMS. Micromachines **11**(1), 7 (2020). https://doi.org/10.3390/mi11010007
2. Yick, J., Mukherjee, B., Ghosal, D.: Wireless sensor network survey. Comput. Netw. **52**(12), 2292–2330 (2008). https://doi.org/10.1016/j.comnet.2008.04.002
3. Ali, A., Ming, Y., Chakraborty, S., Iram, S.: A comprehensive survey on real-time applications of WSN. Future Internet. **9**(4), 77 (2017). https://doi.org/10.3390/fi9040077
4. Mahamuni, C.V.: A military surveillance system based on wireless sensor networks with extended coverage life. In: 2016 International Conference on Global Trends in Signal Processing, Information Computing and Communication (ICGTSPICC), pp. 375–381 (2016). https://doi.org/10.1109/ICGTSPICC.2016.7955331
5. Karray, F., Jmal, M.W., Garcia-Ortiz, A., Abid, M., Obeid, A.M.: A comprehensive survey on wireless sensor node hardware platforms. Comput. Netw. **144**, 89–110 (2018). https://doi.org/10.1016/j.comnet.2018.05.010

6. Sharma, S., Kaur, A.: Survey on wireless sensor network, and its applications. J. Phys. Conf. Ser. **1969**(1), 012042 (2021). https://doi.org/10.1088/1742-6596/1969/1/012042
7. Tomić, I., McCann, J.A.: A survey of potential security issues in existing wireless sensor network protocols. IEEE IoT J. **4**(6), 1910–1923 (2017)
8. Obaidat, M.S., Misra, S.: Principles of Wireless Sensor Networks. Cambridge University Press, Cambridge (2014). https://doi.org/10.1017/CBO9781139030960
9. Singh, M., Kumar, S.: A survey: Ad-hoc on demand distance vector (AODV) protocol. Int. J. Comput. Appl. **161**(1), 38–44 (2017)
10. Mehic, M., Fazio, P., Voznak, M.: Usability of destination-sequenced distance vector routing protocol routes. In: 2019 11th International Congress on Ultra-Modern Telecommunications and Control Systems and Workshops (ICUMT) (2019). https://doi.org/10.1109/icumt48472.2019.8970752
11. Heinzelman, W. R., Chandrakasan, A., Balakrishnan, H.: Energy-efficient communication protocol for wireless microsensor networks. In: Proceedings of the 33rd annual Hawaii International Conference on System Sciences, p. 10. IEEE, January 2000
12. Karp, B. & Kung, H.: GPSR: Greedy perimeter stateless routing for wireless networks. In: Proceedings of the sixth Annual International Conference on Mobile Computing and Networking, pp. 243–254. ACM Press (2000)
13. Lindsey, S., Raghavenda, C.S.: PEGASIS: power efficient gathering in sensor information systems. In: Proceeding of the IEEE Aerospace Conference, Big Sky, Montana, March 2002
14. Younis, O., Fahmy, S.: HEED: a hybrid, energy-efficient, distributed clustering approach for ad hoc sensor networks. IEEE Trans. Mob. Comput. **3**, 366–379 (2004)
15. Qing, L., Qingxin Z., Mingwen W.: Design of a distributed energy-efficient clustering algorithm for heterogeneous wireless sensor networks. In: Computer Communications, vol. 29, issue 12, pp. 2230–2237 Elsevier Science, B.V. Amsterdam, August 2006
16. Elbhiri, B ., Saadane, R., El fldhi, S., Aboutajdine, D.: Developed distributed energy-efficient clustering (DDEEC) for heterogeneous wireless sensor networks. In: I/V Communications and Mobile Network (ISVC), 5th International Symposium, 30 September – 2 October 2010
17. Javaid, N., Qureshi, T.N., Khan, A.H., Iqbal, A., Akhtar, E., Ishfaq, M.: EDDEEC: enhanced developed distributed energy-efficient clustering for heterogeneous wireless sensor networks. In: International Workshop on Body Area Sensor Networks (BASNet- 2013), pp. 1877–0509. Elsevier (2013). https://doi.org/10.1016/j.procs.2013.06.125
18. Saini, P., Sharma, A.K.: E-DEEC- enhanced distributed energy efficient clustering scheme for heterogeneous WSN. In: IEEE 2010 First International Conference on Parallel Distributed and Grid Computing (PDGC), 6 January 2011. https://doi.org/10.1109/PDGC.2010.5679898
19. Smaragdakis, G., Matta, I., Bestavros, A.: SEP: a stable election protocol for clustered heterogeneous wireless sensor networks. In: 2nd International Workshop on Sensor and Actor Network Protocols and Applications, pp. 1–11 (2004). https://arxiv.org/abs/1303.5285
20. Qureshi, T.N., Javaid, N., Khan, A.H., Iqbal, A., Akhtar, E., Ishfaq, M.: BEENISH: balanced energy efficient network integrated super heterogeneous protocol for wireless sensor networks. In: The 4th International Conference on Ambient Systems, Networks and Technologies (ANT 2013), vol. 19, pp. 920–925, 24 June 2013. https://doi.org/10.1016/j.procs.2013.06.126
21. Preethi, A., Pravin, E., Sangeetha, D.: Modified balanced energy efficient network integrated super heterogeneous protocol. In: 2016 Fifth International Conference on Recent Trends in Information Technology. IEEE (2016). https://doi.org/10.1109/ICRTIT.2016.7569562

Check for updates

Towards an Ontology-Based Recommender System for the Vehicle Sales Area

Ngoc Luyen Le[1,2](✉), Marie-Hélène Abel[3], and Philippe Gouspillou[2]

[1] Université de technologie de Compiègne, CNRS, Heudiasyc (Heuristics and Diagnosis of Complex Systems), CS 60 319, 60 203 Compiègne Cedex, France
ngoc-luyen.le@hds.utc.fr

[2] Vivocaz, 8 B Rue de la Gare, 002200 Mercin-et-Vaux, France

[3] Heudiasyc UMR-CNRS 7253, University of Technology of Compiègne, 57 Av. de Landshut, 60200 Compiègne, France

Abstract. In the vehicle sales area, a recommender system (RS) allows vehicle items to be displayed and selected replied on the parameters that are the most relevant to specific users-buyers. In other words, users-buyers can meet with vehicle items suited to their profile, preferences, or interests. Therefore, users will not spend much time searching and retrieving what corresponds to their needs and capacities, as well as against the ever-growing volume of information in the application. In this paper, we realize the analysis of the characteristics of users-buyers and vehicle items in vehicle recommendations. As a result, we propose our approach for building a RS based on ontologies in the vehicle sales area.

Keywords: Recommender system · Ontology · Vehicle recommendation

1 Introduction

In today's era of instant information, more users[1] around the world are becoming more reliant upon Recommender Systems (RSs) to have better advice, suggestions, and inspire them. From e-commercial systems to online advertisements, RSs increasingly influence user's decisions to choose products or services and spend their money, time. Besides, they help users to enhance their experiences on different platforms. These ambitions have been fulfilled with the directly integrated module on their online system and they bring positive evaluations and experiences from users. These systems nudge users to explore new opportunities and suggestions matching their preferences that may never have crossed their minds [20].

In response to this trend, we desire to improve and strengthen user's interactions and experiences by a recommendation engine that will be able to assist users in searching for a vehicle as well as suggesting relevant vehicles to them according to their preferences, history, and needs. The research, analysis, and

[1] We use the term "User" instead of "User-Buyer" in the rest.

L. Troiano et al. (Eds.): ICDLAIR 2021, LNNS 441, pp. 126–136, 2022.
https://doi.org/10.1007/978-3-030-98531-8_13

construction of an effective RS for the purchase/sales of vehicles thus becomes our main objective.

The use of knowledge bases as ontology in the RSs has emerged in terms of the vehicle sales area [16,18]. However, these models have focused on finding and matching users' interest and vehicle items descriptions based on keywords or terms, and semantic relations between these keywords or terms. In this paper, we investigate to go beyond applying ontologies not only to finding and matching data but also to reasoning information and computing similarity between user profiles based on ontologies.

The remainder of this paper is organized as follows. Section 2 studies related works. Section 3 presents our contributions in (i) proposing architecture for the development of an ontology-based recommender system in the vehicle sales area, (ii) inheriting and developing user profile and vehicle ontologies, (iii) calculating similarities between user profile and vehicle items. We present a use case example to illustrate our approach in Sect. 4. Finally, we conclude the paper with remarks for future work in Sect. 5.

2 Related Work

In this section, we introduce the fundamental knowledge about ontologies and recommender systems. In addition, we summarize related work concerning ontologies, RSs, and the use of the ontologies in RSs.

In the context of information science and knowledge sharing, ontology is a form of knowledge representation that consists of a set of concepts within a domain and the relationships between those concepts [6]. Ontology is based on an explicit formal description that allows to shared common understanding of the structure of information, to reuse of domain knowledge, to separate domain knowledge from operational knowledge. The usefulness of ontologies is relied on some reasons as organizing data, improving search, and data integration [2].

One of the purposes of ontology is to capture knowledge about a domain and represent them in a machine-readable and interpretable form. Various ontologies are designed for knowledge representation in the vehicle sales area. More specially, ontologies can be developed to improve semantic search [5]. Or ontologies provided the vocabularies for describing the vehicle such as the Vehicle Sales Ontology (VSO)[2] and the Used Cars Ontology (UCO)[3]. In general, these ontologies are applied to increase the visibility of a vehicle in search engines. On the other sides, the user profile ontology has been widely designed and used in many different applications. The majority of user profile ontologies include set terminology that can describe all characteristics of a user [3].

RS conventionally is an application that attempts to recommend the most relevant items to users by reasoning or predicting the user preferences in an item replied on related information about the users, the items and the interactions between items and users [14]. In general, RSs can be classified into demographic-based RSs, Content-based RSs, Collaboration Filtering-Based (CF-Based) RSs, Knowledge-based RSs, Context-aware RSs, and Hybrid RSs.

[2] http://www.heppnetz.de/ontologies/vso/ns.
[3] http://ontologies.makolab.com/uco/ns.html.

The construction of recommender systems based on using ontologies as a knowledge base has been implemented and experimented within various domains such as course recommendations in e-learning systems [8,17], tour recommendations in tourism applications [11], or product recommendation in e-commercial systems [7]. The use of ontology in the RSs has been developed in the vehicle sales area. For example, the work by [18] has proposed an ontology-based RS that uses an ontology-based hierarchical layered architecture in order to improve semantic searching with understanding hierarchical relations of concepts. *Mu et al.* [16] applied to match between vehicle ontologies and user's interest in order to give the most suitable vehicles corresponding with users' interest.

3 Ontology-Based Vehicle Recommender System

3.1 Needs for Building an Explainable RS in the Vehicle Sales Area

Knowledge bases can be employed to construct an explainable intelligent system. Based on the fundamentals of mathematical logic, systems using a knowledge base can make deductive reasoning provide new information through mathematically proven ways. The construction of an explainable RS in the vehicle domain needs also be carefully considered in the case of helping users better understand the results they receive.

The lack of data directly affects choosing a RS approach as well as relates to the way the evaluation and experiments of the RS. Therefore, gathering and preprocessing data is an indispensable step in the construction of any RS. More specifically, we need to determine which types of data are important and the way how these data will be collected. Besides the organization and management of data are also important factors to be scrutinized.

With RSs, each approach always exists advantages and inconveniences. Using only one approach will benefit about time and cost for the construction of a RS. However, the developed RS will exist shortcomings in some aspects. Therefore, a hybrid system can be a reasonable choice to trade off the limitations of each approach. Although deep neural network models are playing increasingly crucial roles across a wide range of decision-making problems [10]. However, the use of the content-based or the CF-based approaches associated with the deep learning approach perhaps is not the most suitable option in the current condition of data. With special characteristics in the vehicle sales area, the use of knowledge base as ontology in the combination with the content-based and the CF-based approaches in a hybridization design reaches an acceptable solution.

3.2 Development of an Ontology-Based Vehicle Recommender System

The proposed approach is a hybrid recommender system that is combined collaborative filtering, content-based approaches, and ontologies. A combination of both content-based and CF approaches limits shortcomings about cold start or data sparsity problems. Besides using ontology reinforces the accuracy in computing semantic similarity measures about vehicle descriptions, user profiles,

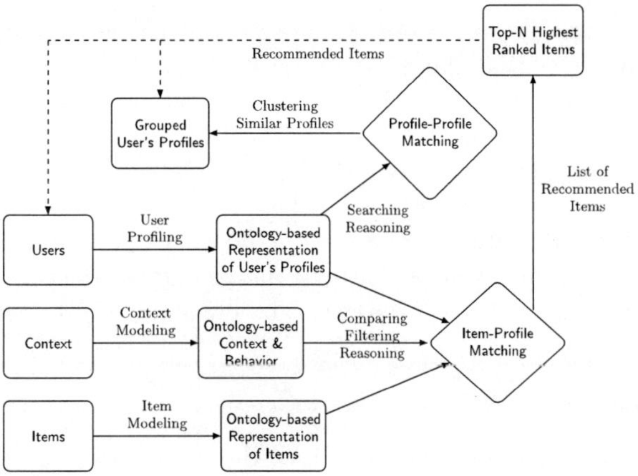

Fig. 1. The recommendation process of the proposed recommender system

and contextual information. As shown in Fig. 1 the process is divided into five steps started by gathering and modeling data to find the recommendation items for each special user, and terminated by filtering and ranking recommendation results as follows:

- Gathering data from different sources and building a knowledge base for users, vehicle items, context, and rating data of items based on ontologies.
- Organizing and mapping different ontologies to link the user profiles, context, and item profile descriptions based on the existing data, mapping the relational database to RDF triples for user profile, item descriptions, contextual information, and interaction data of users for items.
- Filtering and reasoning vehicle items that matched user preferences to reduce computation space.
- Using the content-based approach, calculating the similarity between user preferences and vehicle items by feature vectors of user preferences and vehicle items.
- Using the collaborative filtering approach, calculating the similarity of user recommendation history and similarity between the user profile by using ontology-based semantic similarity.
- Ranking and filtering the list of relevant vehicle items to a user.

3.3 Data Gathering

The limitation of the amount of user and vehicle data is always a challenge in the domain of constructing a recommender system. In our case, the extraction of information will come from both the user and the vehicle. We explore users' information from their profile which is collected from questionnaires and surveys. Besides demographic information, user preferences play also an important role in deeper understanding the intention of users in the decision of purchasing a vehicle. In that regard, we focus user preferences on vehicle color, brand, mileage, budget, the number of seats, and so on.

The vehicle description includes the permanent registration published on open databases such as the Vehicle Identification Number (VIN) which provides manufacturer, portrait of the vehicle, engine size, and type, fuel type, the model year of the vehicle. In addition, the current situation and condition of the vehicle such as the date and expiry of mandatory optional certifications and inspections, non-standard vehicle features, damage or accident information, and others should be collected to have a general view about the vehicle.

3.4 Ontology Construction

This section focuses on constructing knowledge bases with ontology and data sources under triplet sets. The present work allows obtaining two ontologies about user profile and vehicle ontology. Firstly, the user profile ontology model provides a general view of a user and their preferences. We then organize and structure vehicle descriptions with a vehicle ontology.

User Profile Ontology. Many works have constructed user profile ontologies for the applications in various domains from education [1,8], e-commerce [13], to tourism [15]. We develop the User Profile Ontology based on User Profile Ontology (UP)[4]. Besides demographic and social information about a user such as gender, year old, address, occupation, income's range, language, number of kids, married, education level (see Fig. 2), our user profile ontology takes into

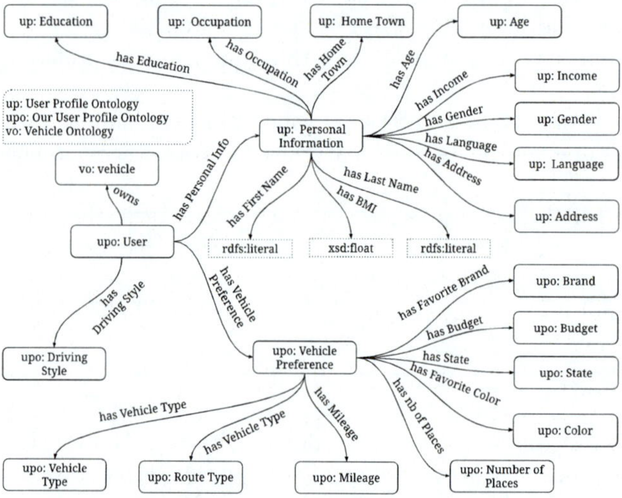

Fig. 2. A visualization of the user profile ontology

[4] http://iot.ee.surrey.ac.uk/citypulse/ontologies/up/up.html.

account psychological information about user preferences that directly influence the consideration and choice of a user to purchase a vehicle. In general, we consider important preferences such as budget, vehicle brand, number of seats, mileage, vehicle type, favorite color, vehicle state, route type.

By using this ontology, the profile of users is organized under triplets in a graph representation. Therefore, a user profile can be formally defined as follows:

$$G_U = \{a_1^u, a_2^u, ..., a_n^u\} \tag{1}$$

where a_i^u denotes the triplet $a_i^u = <subject_i, predicate_i, object_i>$. In other words, the triplet a_i^u can also be expressed as $<resource_i, propertie_i, statement_i>$. For example, "*Henri is interested in blue cars*". This information can be represented in a triplet as $<Henri, hasFavoriteColor, Blue>$. Therefore, a set of triplets of describing a user creates a graph representation where the nodes is equivalent to resources or subject, and arcs denotes named properties.

Vehicle Ontology. Different ontologies have been developed in order to describe and model vehicle information and apply them in several applications. For instance, automobile ontologies have been designed to manage semantic knowledge between the user, the vehicle, and the current driving situation [4,9]. Or an ontology has been proposed to describe tracking, monitoring, and diagnosing problems of an automobile [19]. In our case, the vehicle ontology is based on two ontologies: VSO and UCO. These ontologies provide a standard vocabulary for describing aspects of new and used vehicles for e-commercial applications. Therefore, the vehicle ontology allows a combination of modeling the permanent information such as vehicle identify, license plate number, manufacturer, model, and so on. Besides we can describe status changes of a vehicle such as damage, accident, inspections, mileage (see Fig. 3).

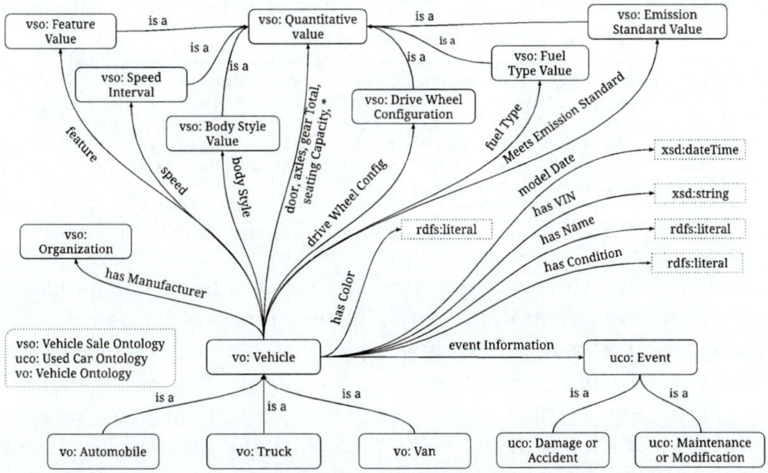

Fig. 3. A visualization of the vehicle ontology

A vehicle description can be represented as a knowledge graph. Similarly, with the user profile, we can define a vehicle description model as follows:

$$G_V = \{a_1^v, a_2^v, ..., a_k^v\} \quad (2)$$

where a_i^v presents the triplet $a_i^v = <subject_i, predicate_i, object_i>$ or $a_i^v = <resource_i, propertie_i, statement_i>$.

When a user makes an interaction with the vehicle by giving a rating, commenting, or adding to a favorite list, we mark these interactions to have an analysis of the intention and behavior of the user and offer relevant recommendations. Therefore, we can further define these interactions as follows:

$$G_I = \{G_U, G_V, r\} \quad (3)$$

where r is the interactive way of the user G_U on the vehicle G_V.

3.5 Semantic Recommendations from Filtering and Reasoning

Ontologies play a key role in the organization and exploitation of knowledge. Ontology provides a common vocabulary and grammar for data as well as semantic relation and description of the data. To develop a knowledge base replied on principles of ontology, different techniques are employed in a combination way from organizing data (e.g., RDF, RDFS, OWL), to querying data (e.g., SPARQL), and to inferring explicit knowledge (e.g., SWRL).

We started with a definition of SWRL rules to inference or calculate information from user profiles. For example, as in the Listing 1.1, we can infer the age of a user from their birthday with two steps: the rule R1 calculates the year born by using the SWRL function *swrl.date*, then the rule R2 uses the *swrl.subtract* function to the year born and the current year to get the age of the user. The rule R3 allows to infer whether a user is a driver or non-driver by comparing the age with the *swrl.greaterThanOrEqual* function.

```
R1 - User(?x), hasBirthday(?x, ?date), swrlb.date(?date,
 ?year, ?month, ?day, ?timezone) => bornInYear(?x, ?year)
R2 - User(?x), bornInyear(?x, ?year), swrlb.subtract(?age,
?nowyear, ?year), thisYear(?nowyear)  => hasAge(?x, ?age)
R3 - User(?x), hasAge(?x,?age), swrlb.greaterThanOrEqual(?age,
18), owns(?x,?vehicle) => isDriver(?x,True)
```

Listing 1.1. Age and driver inference by using the SWRL Rules

In order to make a relevant recommendation, it is important to ensure that the characteristics of vehicle items match the preferences of the user. User profiles and Vehicle descriptions are stored in a triplestore that allows to implementation of efficient querying of a graph-based database. Therefore, we define SPARQL queries to filter and retrieve a list of vehicle descriptions that has the most common characteristics with the user's preference declaration. For example, as in the Listing 1.2, we declare a SPARQL query to retrieve user profiles and vehicle descriptions that have the same favorite color and color of the vehicle.

```
PREFIX vo: <http://vivocaz.fr/vo/ns#>
PREFIX upo: <http://vivocaz.fr/upo/ns#>
SELECT ?user ?vehicle
WHERE {
?user upo:hasPreference ?userPreference .
?userPreference upo:hasFavoriteColor ?color .
?vehicle vo:hasColor ?color .
} ORDER BY ?user LIMIT 10
```

Listing 1.2. A SPARQL query using to search vehicles by the user's favorite color

3.6 Recommendation Computation

In this section, we investigate applying content-based and collaboration filtering approaches to calculating similarities that exist between vehicle items and user preferences as well as between user profiles.

User preference and vehicle item similarity is started with transforming user's interest and vehicle description features to vectors. The user preferences include features as *route type*, *state*, *mileage*, *vehicle type*, *color*, *profile type*, *brand*, *budget* and *number of seats*. These features need to match with vehicle description. A similarity measure is determined to describe the proximity of two vectors. In the simplest case, a cosine similarity can be employed as follows:

$$Sim(G_{U_i}, G_{V_j}) = \frac{\vec{G}_{U_i}.\vec{G}_{V_j}}{\|\vec{G}_{U_i}\|.\|\vec{G}_{V_j}\|} \tag{4}$$

where G_{U_i}, G_{V_j} denote user preferences and vehicle descriptions. Hierarchy ontology similarity can be used to add feature vectors because it helps to add more information based on means of conceptual semantic and lexical relations such as hypernymy, meronymy, and synonym.

User profile similarity allows finding a list of users that have the most similar characteristics with the users who have various interactive activities such as rating, having favorite vehicle item list on the applications. The similarity between the target user and other users is measured by ontology-based semantic similarity $Sim_O(G_{U_i}, G_{U_t})$ based on user profiles and item recommendation history similarity $Sim_H(G_{U_i}, G_{U_t})$ as follows:

$$Sim(G_{U_i}, G_{U_t}) = Sim_O(G_{U_i}, G_{U_t}) + Sim_H(G_{U_i}, G_{U_t}) \tag{5}$$

where G_{U_i} and G_{U_t} denote the user profile i and the target user profile t.

Dive into ontology-based semantic similarity, it can be calculated from feature-based semantic similarities correspondence with using *Object* in a triplet or from information content-based semantic similarities that is equivalent to using *Predicate*. In our case, we use the formula proposed in [12] in order to calculate this index that is defined as follows:

$$Sim_O(G_{U_i}, G_{U_t}) = \sum_{g=1}^{P^{G_{U_i}, G_{U_t}}} \frac{|O_{p_g}^{G_{U_i}} \cap O_{p_g}^{G_{U_t}}| \times \frac{IC(P^{G_{U_i}})}{\sum_{k=1}^{K} IC(P^{G_{U_i}})}}{|O_{p_g}^{G_{U_i}} \cap O_{p_g}^{G_{U_t}}| + \alpha|O_{p_g}^{G_{U_i}} - O_{p_g}^{G_{U_t}}| + \beta|O_{p_g}^{G_{U_t}} - O_{p_g}^{G_{U_i}}|} \quad (6)$$

where $P^{G_{U_i}, G_{U_t}}$ denotes the number of predicates existing in both G_{U_i} and G_{U_t}, p_g is a predicate of G_{U_i} and G_{U_t}, $O_{p_g}^{G_{U_i}}$ indicates a set of qualitative objects related through the predicate p_g, $|O_{p_g}^{G_{U_i}} \cup O_{p_g}^{G_{U_t}}|$ denotes the number of common objects between G_{U_i} and G_{U_t} through the predicate p_g, $|O_{p_g}^{G_{U_i}} - O_{p_g}^{G_{U_t}}|$ represents the number of different objects of $O_{p_g}^{G_{U_t}}$ in $O_{p_g}^{G_{U_i}}$, $IC(P^{G_{U_i}})$ is the amount of information provided by the predicates that G_{U_i} contains, $\sum_{k=1}^{K} IC(P^{G_{U_i}})$ is the sum amount of information provided by the predicates in G_{U_i}, and $\alpha, \beta (\geq 0)$ expresses the weights of G_{U_i} and G_{U_t}.

With the calculation of user's recommendation history similarity, we use also cosine similarity algorithm which is defined as follows:

$$Sim_H(G_{U_i}, G_{U_t}) = \frac{\sum_{k \in R} G_{U_i}.G_{U_t}}{\sqrt{\sum_{k \in R} G_{U_i}^2} . \sqrt{\sum_{k \in R} G_{U_t}^2}} \quad (7)$$

where R is the set of vehicle items that have been interacted by user G_{U_i}, G_{U_t}.

4 Use Case Example

We illustrate a use case example in order to show how our approach works. From the ontologies, we collect maximal information about the user and vehicles and represent them in triplets. For example, we have the triplets of the user Henri *<Henri, hasPersonalInformation, HenriPI>*, *<HenriPI, hasOccupation, Doctor>*, *<HenriPI, hasBirthday, 04/19/1975>*, *<Henri, hasVehiclePreference, HenriVP>*, *<HenriVP, hasFavoriteColor, Blue>*, *<HenriVP, hasVehicleType, Sedan>*, and *<Henri, owns, TeslaModelS01>*.

By using the rules SWRL, we can infer information around a user. For example, we have information about the age of Henri and identify that the user is a driver or a non-driver based on the rules defined in Listing 1.1. Furthermore, we can filter or query users and vehicles which have common characteristics with Henri by query language SPARQL. More especially, we search vehicles which have the characteristics matched with Henri preferences such as *<RenaultClio01, hasName, Renault Clio 2021>*, *<RenaultClio01, hasColor, Blue>*, *<KiaRio01, hasName, Kia Rio 2021>*, *<KiaRio01, hasColor, Blue>*, *<VolkswagenJettaGLI01, hasName, Volkswagen Jetta GLI 2021>*, *<Volkswagen JettaGLI01, hasColor, Blue>*.

The recommendation vehicle items list is calculated based on the similarity between the user preference and vehicle items. First, the set of triplets achieved from SPARQL queries is used to extract a set of features. These features are then transformed into vectors by using some techniques as Tf-Idf or Bag-of-Words. Finally, the similarity between the user's vehicle preferences and vehicle items is calculated by using the formula 4.

We can add into the recommendation vehicle items list from the calculation of the similarity between user profiles and user's recommendation history by employing the formula 5, 6, 7. For example, a user Louis have information as <*Louis, hasPersonalInformation, LouisPI*>, <*LouisPI, hasOccupation, Doctor*>, <*LouisPI, hasBirthday, 01/12/1985*>, <*Louis, hasVehiclePreference, LouisVP*>, <*LouisVP, hasFavoriteColor, White*>, <*LouisVP, hasVehicleType, Sedan*>, and <*Louis, owns, Peugeot20801*>. Using ontology-based semantic similarity defined in formula 6, we can obtain the similarity between the user Henri and Louis. In addition, we calculate the user's recommendation history with the formula 7. Finally, the ranking score of items gives an order of recommendation items for the user.

5 Conclusion and Perspective

The objective of this study is the development of a RS that can generate personalized vehicle recommendations matched with the needs and capacities of users. Using ontologies of user profiles and vehicle descriptions allows enriching information as well as semantic relationships between users, between vehicles, and between users and vehicles. We compute the similarities by using the score functions in order to rank obtained results. Ontology-based semantic similarity calculation is part important in our approach because we can apply it to our ontology model in various aspects both users and vehicles. Furthermore, we can consider a towards explainable recommendation results in which users can understand why a vehicle item is the right suggestion in their case.

References

1. Abel, M.H.: MEMORAe project: an approach and a platform for learning innovation. Multimed. Tools Appl. 1–15 (2021). https://doi.org/10.1007/s11042-021-11157-8
2. Abiteboul, S., Manolescu, I., Rigaux, P., Rousset, M.C., Senellart, P.: Ontologies, RDF, and OWL, pp. 143–170. Cambridge University Press (2011)
3. El Allioui, Y., El Beqqali, O.: User profile ontology for the personalization approach. Int. J. Comput. Appl. **41**(4), 31–40 (2012)
4. Feld, M., Müller, C.: The automotive ontology: managing knowledge inside the vehicle and sharing it between cars. In: Proceedings of the 3rd International Conference on Automotive User Interfaces and Interactive Vehicular Applications, pp. 79–86 (2011)
5. Greenly, W., Sandeman-Craik, C., Otero, Y., Streit, J.: Case study: contextual search for Volkswagen and the automotive industry. Tribal DDB UK (2011)
6. Gruber, T.R.: A translation approach to portable ontology specifications. Knowl. Acquis. **5**(2), 199–220 (1993)

7. Guia, M., Silva, R.R., Bernardino, J.: A hybrid ontology-based recommendation system in e-commerce. Algorithms **12**(11), 239 (2019)
8. Ibrahim, M.E., Yang, Y., Ndzi, D.L., Yang, G., Al-Maliki, M.: Ontology-based personalized course recommendation framework. IEEE Access **7**, 5180–5199 (2018)
9. Klotz, B., Troncy, R., Wilms, D., Bonnet, C.: VSSo-a vehicle signal and attribute ontology. In: SSN Workshop at ISWC. CEUR Workshop Proceedings (2018)
10. Kurovski, M.: Deep learning for recommender systems: joint learning of similarly and preference. Master's thesis (2017)
11. Lee, C.I., Hsia, T.C., Hsu, H.C., Lin, J.Y.: Ontology-based tourism recommendation system. In: 2017 4th International Conference on Industrial Engineering and Applications (ICIEA), pp. 376–379. IEEE (2017)
12. Li, S., Abel, M.H., Negre, E.: Ontology-based semantic similarity in generating context-aware collaborator recommendations. In: 2021 IEEE 24th International Conference on Computer Supported Cooperative Work in Design (CSCWD), pp. 751–756. IEEE (2021)
13. Liu, W., Jin, F., Zhang, X.: Ontology-based user modeling for e-commerce system. In: 2008 Third International Conference on Pervasive Computing and Applications, vol. 1, pp. 260–263. IEEE (2008)
14. Lu, J., Wu, D., Mao, M., Wang, W., Zhang, G.: Recommender system application developments: a survey. Decis. Support Syst. **74**, 12–32 (2015)
15. Martínez-García, M., Valls, A., Moreno, A.: Inferring preferences in ontology-based recommender systems using WOWA. J. Intell. Inf. Syst. **52**(2), 393–423 (2019). https://doi.org/10.1007/s10844-018-0532-5
16. Mu, X., Chen, Y., Li, N., Jiang, J.: Modeling of personalized recommendation system based on ontology. In: 2009 International Conference on Management and Service Science, pp. 1–3. IEEE (2009)
17. Obeid, C., Lahoud, I., El Khoury, H., Champin, P.A.: Ontology-based recommender system in higher education. In: Companion Proceedings of the The Web Conference 2018, pp. 1031–1034 (2018)
18. de Paiva, F.A.P., Costa, J.A.F., Silva, C.R.M.: An ontology-based recommender system architecture for semantic searches in vehicles sales portals. In: Polycarpou, M., de Carvalho, A.C.P.L.F., Pan, J.-S., Woźniak, M., Quintian, H., Corchado, E. (eds.) HAIS 2014. LNCS (LNAI), vol. 8480, pp. 537–548. Springer, Cham (2014). https://doi.org/10.1007/978-3-319-07617-1_47
19. Reymonet, A., Thomas, J., Aussenac-Gilles, N.: Ontology based information retrieval: an application to automotive diagnosis. In: International Workshop on Principles of Diagnosis (DX 2009), pp. 9–14. Linköping University, Institut of Technology (2009)
20. Schrage, M.: Recommendation Engines. MIT Press, Cambridge (2020)

Dominance Relation Based Ranking Procedure for Automated Reverse Auctions

Imène Brigui[1(✉)] and Salem Chakhar[2,3]

[1] Emlyon Business School, Écully, France
brigui-chtioui@em-lyon.com
[2] Portsmouth Business School, University of Portsmouth, Portsmouth, UK
salem.chakhar@port.ac.uk
[3] Centre for Operational Research and Logistics, University of Portsmouth, Portsmouth, UK

Abstract. Most of existing multicriteria auctions systems are very demanding in terms of preference parameters, which requires a considerable cognitive effort from the buyer. In this paper, we propose an auction approach that relies on a combination between Agent systems and dominance relation-based ranking procedure, allowing a buyer agent to manage automatic preference-based English reverse and adaptive auctions. The ranking procedure is based on the dominance relation, which is the most elementary preference information. Within the proposed approach, no aggregation is needed and the ranking procedure will not ask any additional information from the buyer, permitting thus to considerably reduce his/her cognitive effort.

Keywords: Multicriteria english auctions · Multiagent systems · Ranking procedure

1 Introduction

Multicriteria negotiation (e.g. [1]) and especially on multicriteria English auctions (e.g. [2,3]) can be considered as a logical extension of classical auction theory. Their aim is go beyond price-based negotiation to include other attributes in the negotiation process. In reverse English auctions, the process starts when a buyer agent sends a request for an item to all interested supplier agents. Suppliers evaluate the request and send out acceptable bids or withdraw from the competition. The buyer waits to receive all the replies from suppliers, and then selects the best bid and sends out a counterproposal based on the current best-bid utility. The process goes on until all but one supplier quit, and ends with a commitment to the last supplier(s). The scoring function of the buyer may be transmitted to the suppliers.

In iterative auction mechanisms, the process is based on multiple, time-discrete rounds of bidding. The auctioneer provides counterproposals to bidders

L. Troiano et al. (Eds.): ICDLAIR 2021, LNNS 441, pp. 137–145, 2022.
https://doi.org/10.1007/978-3-030-98531-8_14

to help them decide how they can improve their bids. Whereas, in price auctions, only price could be improved by the supplier (a win-lose situation), multicriteria auctions allow the overall bid value to be increased, considering all negotiable attributes.

Most of existing multicriteria based auctions systems need the specification of several preference parameters. This necessitates a considerable cognitive effort from the seller, especially for those with no or limited knowledge about multicriteria analysis. In this paper, we propose an auction approach that relies on a combination between Agent systems and dominance relation-based ranking procedure, allowing a buyer agent to manage automatic preference-based English reverse and adaptive auctions. The ranking procedure is based on the most elementary preference information, which is the dominance relation. The ranking procedure will not ask any additional information from the buyer, permitting thus to considerably reduce the needed cognitive effort from the buyer.

The rest of the paper is organized as follows. Section 2 briefly introduces the principles of proposed auction approach. Section 3 details the ranking procedure. Section 5 concludes the paper.

2 Principles of Auction Approach

The proposed auction approach relies on a combination between Agent systems and dominance relation-based ranking procedure and allows a buyer agent to manage automatic preference-based English reverse and adaptive auctions. We first introduce the following notations. In the rest of this paper, we denote by U the set of potential bids, where any bid is characterized by a set C of attributes.

The evaluation of a bid $x \in U$ with respect to attribute $q \in C$ is denoted by $f(x, q)$. The domains of attributes are supposed to be ordered according to a decreasing or increasing preference. Three types of attributes are considered:

- gain-type: the more, the better: if the preference is increasing with the attribute values.
- cost-type: the less, the better: if the preference is decreasing with the attribute values.
- nominal: when the two first cases are not relevant.

2.1 Negotiation Algorithm

The main assumptions of negotiation algorithm are:

- Each auction deals with a single unit of an item, a fixed set of sellers, and has a fixed deadline.
- Each seller proposes at most one bid at each round.
- The seller who proposed the best bid at the current round is not called upon for the next round.
- At any round, if none of the called upon sellers has provided a bid before the round time limit, then none of them owns a bid satisfying the request constraint.

We assume that the auction starts with the name of the item to be purchased and the set of seller agents that provide the item.

The negotiation buyer agent's algorithm is decomposed into four steps as follows:

(1) **Information collection**. The buyer agent collects the buyer's preferences (the aspiration levels and the reservation levels) and the closing time of the auction.
(2) **Call for proposition**. The buyer agent defines the time duration of a negotiation round and sends to the n seller agents a call for propose message with the closing time as arguments.
(3) **Auction loop**. The buyer agent repeat the following operations until the end of the auction, i.e. the set of competitive sellers is empty or the closing time is reached:
    a. evaluates the received bids and selects the best one as the reference bid for the next round and states the corresponding seller as active.
    b. sends his new request to the remaining seller agents except for the active seller.
    c. waits for the seller bids (receiving propose messages and/or abort messages) and collects all the bids.
(4) **Auction end**. The auction fails if there is no seller in competition. Otherwise, the auction succeeds and the winner is chosen as follows: if the closing time is reached, then the buyer agent selects the best bid as the winning bid, sends an accept message to the winner and a reject message to the other sellers. Otherwise, the best bid of the previous round is the winning bid and the buyer agent sends an accept message to the corresponding seller.

At each round t, the seller agent uses a ranking procedure to rank order the set of bids considered during round t. The ranking procedure embedded within the proposed system is detailed in Sect. 3.

2.2 Bidding Process

This iterative step is the core of the algorithm. At iteration t, the system uses the ranking procedure to rank order the received bids. Agent buyer selects the best bid $best^t$ and sends out a counter-proposal to all buyer except the one proposed $best^t$.

At each round t of an English auction, the buyer agent collects all the bids, selects the best one as the reference bid for the next round and formulates the counterproposal, which is defined as the $best^t$.

3 Ranking Procedure

A ranking is a total order, possibly with ties. The latter are represented by an equivalence relation over U, so the total order is on those equivalence classes. For notational convenience, we define a ranking function r as $r : U \times U \rightarrow \{>, =, <\}$, deciding for any pair of objects whether the first is more likely (>), equally likely (=), or less likely (<) to be more preferred than the second.

In this paper, the ranking function is simply defined through a scoring function. The latter is based on the dominance relation, which is the most elementary preference information.

3.1 Dominance Relation

A *dominance relation* Δ associated with attributes in C is defined for each pair of objects x and y as follows:

$$x \Delta y \Leftrightarrow f(x, q) \geq f(y, q), \forall q \in C. \tag{1}$$

In this definition, the symbol "$\geq$" should be replaced with "$\leq$" for attributes which are ordered according to decreasing preference and by "$=$" for those with no preference order (as for example for nominal or symbolic attributes).

It is important to mention that the definition given above represents the weak version of the dominance relation since all the inequality in the definition above are large. Clearly, this version of dominance relation is reflexive (i.e. $x \Delta x$, $x \in U$) and transitive (i.e., if $x \Delta y$ and $y \Delta z$, then $x \Delta z$, $\forall x, y, z \in U$) (but not complete). Consequently, it defines a partial preorder on set U. The strict version of the dominance relation requires at least one strict inequality in the above given definition. The strict dominance relation is no longer reflexive but still transitive.

To each object $x \in U$, we associate two sets:

- $\Delta^+(x) = \{y \in U : y \Delta x\}$.
- $\Delta^-(x) = \{y \in U : x \Delta y\}$.

The *dominating set* $\Delta^+(x)$ contains the objects that dominate x while the *dominated set* $\Delta^-(x)$ contains the objects dominated by x.

3.2 Partial Scores

We associate to each decision object x two partial scores. The first partial score relies on dominating $\Delta^+(\cdot)$ and dominated $\Delta^-(\cdot)$ sets defined earlier. Let first define the numbers $q^+(x)$ and $q^-(x)$ as follows:

$$q^+(x) = \frac{|\Delta^+(x)|^2}{n_l^2}. \tag{2}$$

$$q^-(x) = \frac{|\Delta^-(x)|^2}{n_l^2}. \tag{3}$$

The measures $q^+(x)$ and $q^-(x)$ can be seen as the power of decision objects that dominating and dominated by decision object x, respectively. The use of the power function permits to ensure that the value of $q^+(x)$ and $q^-(x)$ vary exponentially with the number of objects in $\Delta^+(x)$ and $\Delta^-(x)$, respectively. The first partial score is then defined as follows:

$$s_1(x) = \frac{q^+(x)}{q^+(x) + q^-(x)}, \quad \forall x \in U. \tag{4}$$

However, this version will not solve completely the problem of ties since any two decision objects x and y verifying $\Delta^+(x) = \Delta^+(y)$ and $\Delta^-(x) = \Delta^-(y)$ will have the same score. To further reduce/remove the problem of ties, we can exploit partial dominance relations, i.e., dominance relations with respect to single attribute. The partial dominance relation with respect to criterion $q \in C$ is defined as follows for any pair of objects x and y, $x, y \in U$:

$$\delta_q(x, y) = \begin{cases} 1, \text{ if } f(x, q) \geq f(y, q) \\ 0, \text{ otherwise.} \end{cases} \tag{5}$$

The operator "$\geq$" should be replace by "$\leq$" for cost attributes and with "$=$" for nominal attributes. Then, we define two additional measures, denoted by $p^+(x)$ and $p^-(x)$, respectively, as follows:

$$p^+(x) = \sum_{y \in U \wedge y \neq x} |\sum_{q \in C} \delta_q(y, x)|^2. \tag{6}$$

$$p^-(x) = \sum_{y \in U \wedge y \neq x} |\sum_{q \in C} \delta_q(x, y)|^2. \tag{7}$$

The use of the power function enables us to ensure that the value of $p^+(x)$ and $p^-(x)$ vary exponentially with the number of objects in $\delta(\cdot, \cdot)$. The second partial score is then defined as follows:

$$s_2(x) = \frac{p^+(x)}{p^+(x) + p^-(x)}, \quad \forall x \in U. \tag{8}$$

3.3 Scoring Function

The scoring function is defined as follows:

$$s : U \rightarrow \mathbb{R}^+ \tag{9}$$

$$x \rightarrow s(x) \tag{10}$$

with

$$s(x) = \sum_{i=1}^{2} s_i(x). \tag{11}$$

It is important to notice that ties still may exist. This holds when: (i) two different objects have exactly the same scores on the attributes, (ii) two objects such that $\sum_q \delta_q(x, y) = \sum_q \delta_g(y, x)$ or two objects that differ only with respect to nominal attributes. The first case is consistent since the two objects should naturally have the same rank. The next two cases may be handled by considering the difference between the evaluations of objects with respect to attributes. This is outside the scope of this paper.

3.4 Ranking Function

Let first mention that to be useful, the scoring function needs to ensure that the scores of any bid x must be strictly higher to the score of bid y for all $x, y \in U$ and $x\Delta y$. The scoring function $s(.)$ defined earlier meets this requirement, as stated by the following theorem.

Theorem 1. *Let $x, y \in U$ with $x\Delta y$. Then, the following holds:*

$$s(x) > s(y). \tag{12}$$

Proof. Trivial as the dominance relation is transitive by definition.

The ranking function is then simply defined using the scoring function specified above:

$$r(x, y) \begin{cases} >, s(x) > s(y), \\ =, s(x) = s(y), \\ <, s(x) < s(y). \end{cases} \tag{13}$$

4 Illustrative Example

For illustration purpose, we consider the case a buyer looking for new though Auction process. Let also assume that the list of attributes are as shown in Table 1. Let us assume that the buyer received the list of bids given in Table 2. In what follows, we illustrate the ranking procedure to rank order the list of bids.

Table 1. List of attributes

Code	Name	Data Type	Preference
A1	Price	Continuous	Cost
A2	Swept volume	Continuous	Cost
A3	Fuel consumption	Continuous	Gain
A4	Horsepower	Continuous	Cost
A5	Torque	Continuous	Cost

Table 2. Initial list of bids

Bid x_k	A1	A2	A3	A4	A5
1	107.9	2495	20.6	161	22.7
2	140.9	2495	20.6	161	22.7
3	334	2144	22.6	205	52
4	552	3499	14.6	307	38.7
5	587	4396	7.3	556	70.3
6	240	648	92.9	35	6.61
7	990	1500	48.6	232	33.6

The rank procedure is then automatically applied on list of bids. The dominance relation applied on the list of bids is given in Table 3. To compute $q^+(x_k)$ and $q^-(x_k)$ of bid x_k ($k = 1, \ldots, 7$), we need first to compute sets $\Delta^+(x_k)$ and $\Delta^-(x_k)$. The results are shown in Table 4.

Table 3. Dominance matrix

	1	2	3	4	5	6	7
1	1	1	0	1	1	0	0
2	0	1	0	1	1	0	0
3	0	0	1	0	1	0	0
4	0	0	0	1	1	0	0
5	0	0	0	0	1	0	0
6	0	0	1	1	1	1	1
7	0	0	0	0	0	0	1

Table 4. Dominated $\Delta^+(x_k)$ and dominating $\Delta^+(x_k)$ sets

x_k	$\Delta^+(x_k)$	$\Delta^+(x_k)$
1	1, 2, 4, 5	1
2	2, 4, 5	1, 2
3	3, 5	3, 6
4	4, 5	1, 2, 4, 6
5	5	1, 2, 3, 4, 5, 6
6	3, 4, 5, 6, 7	6
7	7	6, 7

Then, the values of $q^{+}(x_k)$, $q^{-}(x_k)$ and partial scores $s_1(x_k)$ can then be computed through Equations (2), (3) and (4), respectively. The results are summarised in Table 5.

Table 5. Values of $q^{+}(x_k)$, $q^{-}(x_k)$ and partial scores $s_1(x_k)$

x_k	$q^{+}(x_k)$	$q^{-}(x_k)$	s_1
1	0.32653	0.02041	0.94118
2	0.18367	0.08163	0.69231
3	0.08163	0.08163	0.5
4	0.08163	0.32653	0.2
5	0.02041	0.73469	0.02703
6	0.51020	0.02041	0.96154
7	0.02041	0.08163	0.2

Equations (6), (7) and (8) are then used to calculate $p^{+}(x_k)$, $p^{-}(x_k)$ and partial scores $s_2(x_k)$. The results are summarised in Table 6.

Table 6. Values of $p^{+}(x_k)$, $p^{-}(x_k)$ and partial scores $s_2(x_k)$

x_k	$p^{+}(x_k)$	$p^{-}(x_k)$	s_2
1	698	812	0.46225
2	625	865	0.41946
3	193	657	0.22706
4	92	253	0.26667
5	147	192	0.43363
6	64	49	0.56637
7	302	530	0.36298

Finally, the overall score $s(x_k)$ is computed using Equation (12). The obtained results along with final ranking of bids is given in Table 7. As shown in this table, bid #6 is ranked first and should be selected by the buyer.

Table 7. Overall scores and ranks of bids

x_k	$s(x_k)$	Rank of x_k
1	1.40343	2
2	1.11177	3
3	0.72706	4
4	0.46667	6
5	0.46066	7
6	1.52791	1
7	0.56298	5

5 Conclusion

The paper proposed an auction approach that combines Agent systems and dominance relation-based ranking procedure. As the latter is build upon the dominance relation, it reduces considerably cognitive effort required from the buyer. This means that no additional information are requested from the buyer. This is an important feature of the proposed system.

As mentioned in Sect. 3.3, ties still can exist in some specific cases. To address this issue, we intent to extend the scoring function by adding a third partial score relying on the difference between the evaluations of objects with respect to attributes. Future work will also consider the application of the proposed approach using simulated and real-life datasets. Additionally, we intend to investigate the performance and computational behavior of proposed approach.

References

1. Bellantuono, N., Ettorre, D., Kersten, G., Pontrandolfo, P.: Multi-attribute auction and negotiation for e-procurement of logistics. Group Decis. Negot. **23**, 421–441 (2014)
2. Bellosta, M.-J., Kornman, S., Vanderpooten, D.: Preference-based English reverse auctions. Artif. Intell. **175**(7), 1449–1467 (2011)
3. Pham, L., Teich, J., Wallenius, H., Wallenius, J.: Multi-attribute online reverse auctions: recent research trends. Eur. J. Oper. Res. **242**(1), 1–9 (2015)

Predicting Business Failure Using Neural Networks: An Empirical Comparison with Statistical Methods and Data Mining Method

Yaser Allozi(✉) and Maysam Abbod

Department of Electronic and Computer Engineering, Brunel University London, Uxbridge, UK
{Yaser.allozi,Maysam.Abbod}@brunel.ac.uk

Abstract. Predicting business insolvency is considered one of the main supportive sources of information for decision making by financial institutions, investors, creditor, and other participants in the business. The financial reporting system provide relevant information that could be used to assess the financial position of the firms to help users making right decision about the firm. It is become crucial to have an early prediction model that provide accurate assurance for users about the health situation of the business. Recent studies focused on using data mining tools instead of traditional statistical methods for developing models to classify firm into healthy or failed according to information from its financial ratios extracted from their financial reports. In this study, we develop insolvency perdition models based on three modelling techniques, named, logistic regression (LR), (as a statistical widely used method) and artificial neural network (ANN) and support vector machine (SVM) (as a data mining modelling tools). Results of the study revealed the outperformance of SVM in predicting financial insolvency based on performance measurements.

Keywords: Business insolvency · Logistic regression (LR) · Artificial neural networks (ANN) · Support vector machine (SVM)

1 Introduction

Nowadays competitions among business institutions to attain their market share in order to survive has become the main concern of all participant in the business industry. Many factors such as weak corporate governance, lack of innovation, and weak performance of business management in a market full of uncertainty can lead to business failure. However, the financial reporting system consider as the major indicator used by participants to get information about the businesses and reveal the possible failure that might accore through the process of financial analysis.

The financial statement reporting system defined as the process in which the organization deliver its financial information to different users such as suppliers, creditors, investors, and others using five essential financial statements: statement of financial position, income statement, cash flow statement, change in equity statement, and statement

L. Troiano et al. (Eds.): ICDLAIR 2021, LNNS 441, pp. 146–156, 2022.
https://doi.org/10.1007/978-3-030-98531-8_15

of comprehensive income. These Financial statements should provide relevant, reliable and faithful representation of financial information to users to support their decision-making process. According to [17] the Relevance of financial statement is defined as the capability of the delivered financial information to make difference for users' decisions making. Faithful representation means that the financial information represents the real economic situation in the market. It means that the financial information has confirmatory and predictive value that is reflected on the real market Therefore, many researcher has tested the value relevant of the financial information, and they normally define it as value relevant [17, 22, 23].

Financial analysts use financial ratios, extracted from the financial statements, alongside financial statement figures to interpret firm's profitability, performance, operational capabilities, capital structure, and solvency. These ratios and figures have been used by authors to build prediction models in various accounting area such as financial statement fraud, financial distress, credit rating, earnings prediction, and auditing.

Ou and Penman [21] conducted the first study in this area where they used financial ratios to develop an earning prediction model based on statistical method. The aim of the model is to predict earnings increase in the subsequent fiscal year of the business. Attribute selection was used to define the relationship between variables in order to select best variable for modelling an accurate prediction model. However, statistical methods also used for modelling business failure models, which are aligned with assessing the potential risk of business failure through an early prediction of firm distress using financial ratios. The models are developed to classify firms in two group either 'good' or 'failed' with respect to their financial information.

The large amount of information on firms' financial data attracts the attentions of many researchers to apply data mining tools as modelling methods in order to improve the prediction accuracy of firms' situation. It is been concluded from the literature that machine learning techniques outperform statistical method in terms of prediction accuracy of business insolvency and financial distress especially for non-linear relationship between variables. However, despite of the impressive results achieved by researcher on the field and the ability of data mining tools to classify using financial data, there is still a call by academic for using large data sets and more attributes to develop models and increase its accuracy.

The purpose of this study is to clarify the potential capability of data mining as a classification tool in accounting and finance area to predict business failure. In this study, an attempt to develop an accurate model using variables from financial ratios and accounting items to predict business failure by using data mining techniques. This study we use big data of financial information to develop a model capable to predict business failure.

2 Related Works

Financial variable has been used for developing prediction models in many different areas such as financial distress, financial fraud detection, auditing, and other area in the accounting field. Therefore, developing an early prediction model of business insolvency is demanded by financial statement users such as financial institutions, auditor, investors, creditor, financial analysts, and other users, due to the supportive information it provides for decision making. The research tendency in the literature is going towards developing machine learning and data mining tools for modelling, since it achieved higher performance in comparison with traditional statistical methods.

In the auditing area, auditors provide their opinion on the possibility of firms to fail through their report on its going concern opinion. A study by [19] conducted as a study to investigate the capability of data mining to predict ongoing concern decisions based on financial ratios of firms that have received an ongoing concern. They developed a prediction models based on both statistical method (logistic regression), as a widely used method in the field, and different data mining techniques. Their result show that data mining techniques outperformed logistic regression in term of classifying average accuracy rate. However, their research includes only four financial ratios where more productive and relevant variables can be used to enhance the prediction accuracy of the model.

Barboza, et al. [24] compared the utilisation performance of both traditional method (discriminant analysis, logistic regression, and neural network) and machine learning (support vector machine, bagging, boosting, and random forest) in developing bankruptcy prediction model. Their model is based on eleven financial variables consisting of five Altman's Z-score variables and another six complementary financial indicators related to 10,000 US firm-year observations. Their results show on average, 10% increase in accuracy rate of machine learning models in comparison with traditional models. The study supports the debate about the superiority of machine learning performance over traditional statistical methods in bankruptcy prediction modelling [5] and [10]. However, in terms of machine learning, SVM method did not get the highest accuracy rate which contradict other study of [13] and [12] and corroborates [9, 11] and [8].

In an attempt to enhance model prediction accuracy [11] combine seven different categories of financial ratios variables with variables related to five different categories of corporate governance to develop a bankruptcy model using data mining techniques based on a data from Taiwan. They used data consist of 239 bankrupt and 239 non bankrupt firm with 95 financial ratios and 95 corporate governance ratios as an input data for modelling. Their result shows better prediction performance for those models with corporate governance indicators than other model with financial ratios alone. Moreover, the result show that model based on support vector machine SVM outperformed other data mining techniques in terms of prediction accuracy.

Sayari and Mugan [16] conducted an investigation study on the capability of financial ratios to reveal the uncertainty level of various industries to determine the set of industries characteristics. They used 51 financial ratios extracted from the literature to develop an industry specific financial distress prediction model using logistic regression method on data from S&P 1500 firms. Their findings show that financial ratios as an industry informative content varies among different industry groups and the model accurately predict financial distress. However, the classification accuracy rates diverge across different industry groups.

Xx [15] conduct an in-depth review study on 49 journal articles published during the period from 2010 to 2015 related to bankruptcy prediction model development in order to provide a platform for researcher to select best tools for modelling. They used Meta-analysis in combination with 'finding summery' tables and statistical charts in order to discover the most useful tools used to develop the bankruptcy prediction model. The analysis summery indicate that there is eight popular and promising tools (two statistical tools and six artificial intelligence tools) have been used by authors in this research area according to 13 keys criteria. However, they concluded that each tool has its strength and weaknesses where there is no singular tool outperform other tools in relation to all identified criteria in the study.

Amani and Fadlalla [14] explored the applications of data mining on the accounting field through a systematic review of the topic in previous literature. Their research methodology consists of seven steps to capture most related literature from different data resources. 209 data mining and accounting related papers was selected to be studied in order to be presented in a structurally logical and thematically coherent approach. Their result shows that 82% of the research used data mining applications for predictive goals, and 11%, and 7% for descriptive and prescriptive goals respectively. And in term of research topics, 64% of the researches are related to accounting assurance and compliance. Managerial accounting consists of 25% and financial accounting and accounting information system topics consist of 11% of the researches. However, the reviewed researches showed that the main task of using data mining techniques is for classification followed by estimation purposes.

These studies focused on the application of data mining modelling methods in order to achieve the highest predictive accuracy. Therefore, it can be observed that vast majority algorithms such as ANN, SVM, and DT has been used as the base method for classifying data. However, none of these studies used large size of data for modelling.

For this reason, we propose a comparison study between three models, logistic regression as a statistical approach, and ANN, SVM, and DT, as a machine learning techniques to investigate in their performance in predicting business failure using large size of data.

3 Modelling Methods

A. Logistic Regression
LR has been used widely by academics as a statistical technique for solving classification problems. It is used for binary classification where the outcome represented by zero or one depending on the probability of event to occurs based on the independent variables in the model [25]. Therefore, the probability of the event is conditional to the values of the independent variables.

Hence the probability of business to fail have to be binary as (healthy, 1 or fail, 0) based on the relationship of the independent variables. LR formula is represented in the following equations:

$$\mathrm{Log}\left[\mathrm{p}\left(1-\mathrm{p}\right)\right] = \beta_0 + \beta_1 X_1 + \beta_2 X_2 + \cdots + \beta_n X_n \quad (1)$$

where p is the probability of the dependent variable to occur (0, 1), β_0 represent intercept, and $\beta 1 - \beta n$ represents the coefficient of each independent variable $X1 - Xn$ in the model, and $[p(1 - p)]$ is the dependent variable which equal to the logarithm of ratio of two probability outcomes. The task of LR in business insolvency is to determine the conditional probability of input (the characteristics of financial variables) related to a certain class [26].

It is been observed that classification accuracy rate decrease when the relationship between variables are nonlinear and considered as a limitation of using LR in insolvency prediction. Therefore, data mining tools are introduced in order to overcome this limitation.

B. Data Mining

Because of the vast development on the information communication technology, a Large amount of Data from different sources has generated and normally stored based on specific own structure in different geological locations. Applying data mining techniques on these data, which tend to be irrelevant to each other, is useful to discover patterns between it to generate critical important knowledge for decision-making. Data mining is defined as the process of extracting information from different data sources and transform it into a comprehensible structure for different further use, using computer science, machine learning, database systems, and statistics. However, data classifying is defined as a method used to classify data based on patterns between variables, which in return reduce time consuming, increase accuracy, and reduce cost of misclassification. The following introduce data mining tools used for classification in the field of business insolvency prediction.

1) Artificial Neural Networks

Artificial neural network (ANN) is as a computational or a mathematical modelling tool of non-linear data, constructed based on an emulation of the biological neural system (human brain function). The concept of ANN is a mimic of the neural system in which a group of artificial neurons are interconnected to process information using a connectionist approach to computation and to learn and adapt form historic data [1]. ANN has been used by academics for modelling business insolvency and financial distress as an alternative to traditional statistical methods.

An ANN has to be configured in a way to achieve the desired set of output from the application of a set of inputs.as show in Fig. 1, ANN topology consists of three layers: input, hidden, and output layers. The structure of business insolvency model starts by feeding financial variables, as an attribute of each firm, to the input layer to process them, and then they are processed more in the hidden layer. Finally, after thorough processing of the data the values sent to the output to give an answer weather a firm is healthy or going to fail. However, weights are assigned to each attribute to calculate the output based on its relative importance. Weights will be adjusted based on supervised learning rule in which input and output data are fed to the model to be used in training the model. All of the weights are summed together using transfer function (i.e. sigmoid, tansig) in order to predict output. The process of adjusting weights will be repeated iteratively to minimize the error between predicted and actual output [25].

The core advantage of ANN is it capability to find the relationship between variables and deal with nonlinearity and choose among the most important predictable attributes in the model. Moreover, it has the capability to model using incomplete data sets with missing and noisy data with no previous assumption of data distribution that allow it to

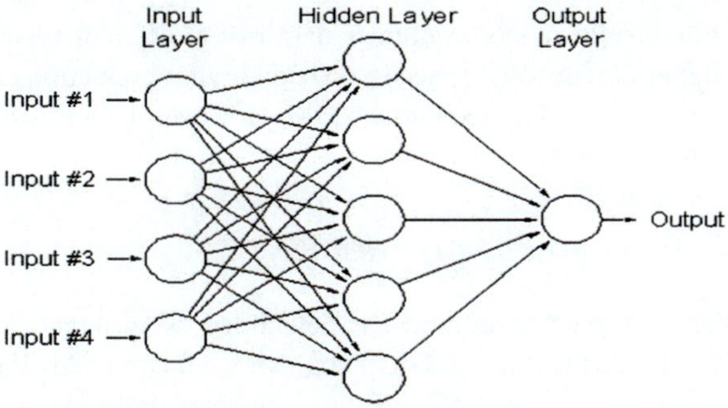

Fig. 1. ANN topology [2].

recognise complex pattern between attributes [29]. However, long training process and lack of theory consider as a drawback of ANN modelling technique.

In this study, we used feed forward backpropagation as an ANN modelling method for classifying fault firms.

2) Support Vector Machine
Support Vector Machine considered as a supervised learning modelling technique derived from statistical learning theory which is associated with a learning algorithm used to analyse data in classification and regression [28].

As illustrated in Fig. 2, SVM has a decision surface called the optimal hyperplane and the data points closest to it and the margins (dashed lines) are called support vectors. The support vectors are very important elements used for training the model where the SVM finds the optimal hyperplane separating the data with maximum margin width to fit the data. SVMs is a non-probabilistic binary linear classifier that classify a set of training to one of two categories [28]. However, Kernel function mechanism can be used for mapping process in which the SVM can be adapted to become a nonlinear classifier using nonlinear kernels. Several binary SVM classifiers can be combined to convert SVM into a multiclass classifier.

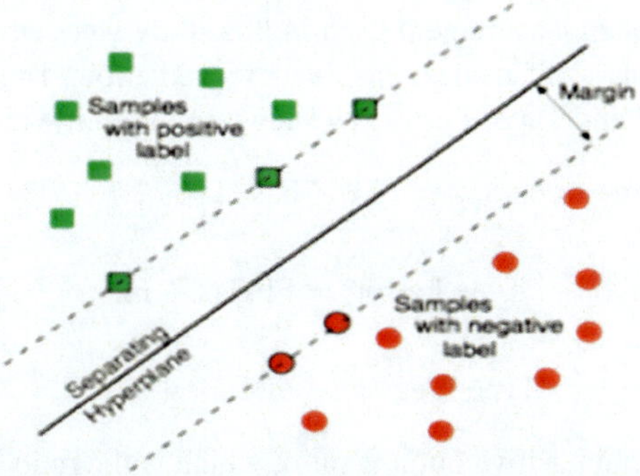

Fig. 2. The basis of support vector machines [28].

SVM has the ability to model nonlinear data using different type of kernels that transfer data in higher dimensional space to provide higher prediction accuracy rates in comparison with other modelling techniques [34]. However, in contrast of ANN, SVM is sensitive towards missing and noisy data.

4 Data Collection and Pre-processing

In this study a solvency prediction model is developed using financial data related to 32267 firms from all industries in the UK. Data extracted from FAME database which is a financial information database of 7 million companies in the UK updated in a daily basis with up to 10 years of data. It provides detailed firms' information including data of financial statements and already calculated financial ratios related to firms' financial performance as shown in Table 1.

Table 1. Description of the datasets used in this study

	Dataset			
	# FIRMS	# Active	# Inactive	# Financial Indicators
UK	32267	19207	13060	37

The Data collected consist of annual ratios and accounting figures related to three consecutive years prior to business failure. This allow to investigate in model accuracy as an early prediction of business failure. 37 financial variables were used in this study as independent indicators of business insolvency. Data was pre-processed by cleaning all missing data using SPSS software. Therefore, only firms with all known value of financial variables was selected for model development.

5 Criteria for Comparing the Models

The model performance measurement used in this study were embraced from previous related work in the area of business insolvency prediction. These measurements are Accuracy rate, Type I and Type II errors as shown in the following formulas:

$$\text{Average Accuracy} = TP + TN/TP + FN + TN + FP \quad (2)$$

$$\text{Type I error} = FP/TN + FP \quad (3)$$

$$\text{Type II error} = FN/TP + FN \quad (4)$$

Where TP stand for true positive outputs ratio, which is the ratio of correctly classified healthy firms. TN refer to the ratio of correctly classified as failed firm. FN and FP measure the ratio of misclassified firms (false positive and false negative) as illustrated in Table 2.

Table 2. Description of the datasets used in the study

		Predicted class (%)	
		Active	Inactive
Actual class (%)	Active	True positive	False negative (Type II error)
	Inactive	False Positive (Type I error)	True Negative

6 Results

This section shows the results of all modelling tools used to build the prediction model based on testing results of performance measurements on 32267 firms in the UK.

First, LR classifier was run on SPSS 26 version using stepwise feature selection method. Only 18 variables out of 37 variables was selected in the equation as the most significant predictable variables based on a confidence level of 95% for year t and $t - 1$ datasets. However, model $t - 2$ selected three more variables of the dataset to the model to obtain the highest prediction accuracy rate as shown in Table 3.

ANN classifier in this study were executed using NPR tool (Neural Pattern Recognition) as a classification tool in MATLAB. Its topology relies on the number of neurons on the hidden layer of the model and in order to figure out the best number, trial and error is used for the selection of neurons ranged from 10 to 30 neurons. Moreover, a predetermined number of epochs was selected for iteration in order to train the mode. The data was splitted into three datasets, as 70% for training, 15% for validation, and 15% for testing. At the end, the best ANN topology was selected based on the highest predictive accuracy accompanied with the lowest Mean Square Error. The classification accuracy of ANN is shown in Table 4.

The SVM classifiers algorithm was also developed on MATLAB classification learning application tool. In order to protect the data against overfitting we used Cross Validation method, in which partitioning the dataset into folds and estimating accuracy based on the average of these folds. Quadratic SVM and produced the highest prediction rates for all data sets.

According to the results on Tables 3 through 5, SVM has outperformed other methods in term of average accuracy rates. The model scored 86,2% on average accuracy for year 1 whereas while ANN achieved 82.95% and LR achieved the lowest accuracy rate of 79%. On the other hand, the prediction accuracy rates of all classifiers has drop down when we predict failure earlier for yesr $t - 1$ and $t - 2$. The results shows a decline in the prediction accuracy rate the earlier we predict failure.

Worth mentioning that ANN classifier has more balanced prediction results represented in it sensitivity and specificity rates in comparison with other two classifiers. The classifier has achieved 83.1% and 82.8% for year 1, respectively. However, the SVM classifier has achieved 87.45 and 85% for the same year showing higher gap between the two rates. In case of the LR results, the gap is higher showing worse results than data mining techniques.

Although the ranking of the classifiers performance is the same for year t − 1 and t − 2, the SVM classifier has better average accuracy in year t − 1. The improvement is resulting form the enhancement in predicting failed firms as can be seen in better type II error rate.

Based in all performance results, SVM has outperform ANN and LR classifiers in correctly predicting and classifying failed firms. This results consist with ability of data mining outperformance over statistical methods. therefore, SVM has the best prediction accuracy for classifying failed firms for the UK dataset.

Table 3. Results of LR

Classifier	Sensitivity	Specificity	Average accuracy rate	Type I error	Type II error
LR (t)	77%	81%	**79%**	19%	23%
LR (t − 1)	76.3%	80.7%	**78.5%**	18.3%	23.7%
LR (t − 2)	75.4%	80.3%	**77.85%**	18.4%	24.6%

Table 4. Results of ANN

Classifier	Sensitivity	Specificity	Average accuracy rate	Type I error	Type II error
ANN (t)	83.1%	82.8%	**82.95%**	17.2%	16.9%
ANN (t − 1)	82%	82.1%	**82.05%**	17.9%	18%
ANN (t − 2)	81.7%	81.6%	**81.65%**	18.4%	18.3%

Table 5. Results of SVM

Classifier	Sensitivity	Specificity	Average accuracy rate	Type I error	Type II error
SVM	87.4%	85%	**86.2%**	15%	12.6%
SVM (t − 1)	88%	84.9%	**86.45%**	15.1%	12%
SVM (t − 2)	85.4%	84%	**84.7%**	16%	14.6%

7 Conclusion and Future Directions

The aim of this study is to investigate on the best prediction method that can be used to build a classification model to predict business failure based on financial indicators in the UK. The proposed model was developed based on collected financial data of 32267 firms in the UK. First dataset was cleaned from missing value, and Then divided into training (to learn the model), validation, and testing (to validate the model) sets as 70%, 15%, and 15% respectively. Machine-learning and statistical classifier were generated to train the model using the training dataset. Finally, the outcomes of the testing dataset were used to compare each method performance based on accuracy rate and Type I and Type II classification error.

For statistical method LR, the classification accuracy rate achieved was the lowest rates for all three-year's datasets in comparison with machine learning results. Therefore, Machine-learning techniques show superiority classification performance and SVM classifier has achieved the classification results. However, ANN accuracy rates relatively show more stability prediction over the years in comparison with LR and SVM accuracy rates.

Future studies in this field could be extended by using non-financial variables related to the market, using different machine learning techniques and ensemble methods such as boosting, bagging, and random forest, and using different data sets from other markets.

References

1. Huang, C., Chen, M., Wang, C.: Credit scoring with a data mining approach based on support vector machines. Expert Syst. Appl. **33**, 847–856 (2007)
2. Tsai, C., Wu, J.: Using neural network ensembles for bankruptcy prediction and credit scoring. Expert Syst. Appl. **34**, 26392649 (2008)
3. Atiya, A.F.: Bankruptcy prediction for credit risk using neural networks: a survey and new results. IEEE Trans. Neural Netw. **12**, 929–935 (2001)
4. Tsai, C.: Combining cluster analysis with classifier ensembles to predict financial distress. Inf. Fusion **16**, 46–58 (2014)
5. West, D., Dellana, S., Qian, J.: Neural network ensemble strategies for financial decision applications. Comput. Oper. Res. **32**, 2543–2559 (2005)
6. Nanni, L., Lumini, A.: An experimental comparison of ensemble of classifiers for bankruptcy prediction and credit scoring. Expert Syst. Appl. **36**, 3028–3033 (2009)
7. Cano, G., et al.: Automatic selection of molecular descriptors using random forest: application to drug discovery. Expert Syst. Appl. **72**, 151–159 (2017)
8. Wang, G., Ma, J., Yang, S.: An improved boosting based on feature selection for corporate bankruptcy prediction. Expert Syst. Appl. **41**(5), 2353–2361 (2014)
9. Yeh, C. C.-C., Chi, D.-J., Lin, Y.-R.: Going-concern prediction using hybrid random forests and rough set approach. Inf. Sci. **254**, 98–110 (2014)
10. Liang, D., Lu, C.-C., Tsai, C.-F., Shih, G.-A.: Financial ratios and corporate governance indicators in bankruptcy prediction: a comprehensive study. Eur. J. Oper. Res. **252**(2), 561–572 (2016)
11. Cleofas-Sánchez, L., García, V., Marqués, A., Sénchez, J.: Financial distress prediction using the hybrid associative memory with translation. Appl. Soft Comput. **44**, 144–152 (2016)
12. Danenas, P., Garsva, G.: Selection of support vector machines based classifiers for credit risk domain. Expert Syst. Appl. **42**(6), 3194–3204 (2015)
13. Amani, F.A., Fadlalla, A.M.: Data mining applications in accounting: a review of the literature and organizing framework. Int. J. Account. Inf. Syst. **24**, 32–58 (2017)
14. Alaka, H.A., et al.: Systematic review of bankruptcy prediction models: towards a framework for tool selection. Expert Syst. Appl. **94**, 164–184 (2018)

15. Sayari, N., Mugan, C.S.: Industry specific financial distress modeling. BRQ Bus. Res. Q. **20**(1), 45–62 (2017)
16. Tsai, C., Hsu, Y., Yen, D.C.: A comparative study of classifier ensembles for bankruptcy prediction. Appl. Soft Comput. **24**, 977–984 (2014). https://doi.org/10.1016/j.asoc.2014.08.047, http://www.sciencedirect.com/science/article/pii/S1568494614004128
17. Barboza, F., Kimura, H., Altman, E.: Machine learning models and bankruptcy prediction. Expert Syst. Appl. **83**, 405–417 (2017)
18. Martens, D., et al.: Predicting going concern opinion with data mining. Decis. Support Syst. **45**(4), 765–777 (2008)
19. Chua, W.F., Taylor, S.L.: The rise and rise of IFRS: an examination of IFRS diffusion. J. Account. Public Policy **27**(6), 462–473 (2008) https://doi.org/10.1016/j.jaccpubpol.2008.09.004, http://www.sciencedirect.com/science/article/pii/S0278425408000938
20. Ou, J.A., Penman, S.H.: Financial statement analysis and the prediction of stock returns. J. Account. Econ. **11**(4), 295–329 (1989). https://doi.org/10.1016/0165-4101(89)90017-7, http://www.sciencedirect.com/science/article/pii/0165410189900177
21. Yip, R.W., Young, D.: Does mandatory IFRS adoption improve information comparability? Account. Rev. **87**(5), 1767–1789 (2012)
22. Clarkson, P.M., et al.: Does it really pay to be green? Determinants and consequences of proactive environmental strategies. J. Account. Public Policy **30**(2), 122–144 (2011)
23. Barth, M.E., Landsman, W.R., Lang, M.H.: International accounting standards and accounting quality. J. Account. Res. **46**(3), 467–498 (2008)
24. Akkoç, S.: An empirical comparison of conventional techniques, neural networks and the three stage hybrid adaptive neuro fuzzy inference system (ANFIS) model for credit scoring analysis: the case of Turkish credit card data. Eur. J. Oper. Res. **222**, 168–178 (2012)
25. Lee, T., Chen, I.: A two-stage hybrid credit scoring model using artificial neural networks and multivariate adaptive regression splines. Expert Syst. Appl. **28**, 743–752 (2005)
26. Ong, C., Huang, J., Tzeng, G.: Building credit scoring models using genetic programming. Expert Syst. Appl. **29**, 41–47 (2005)
27. Singh, Y., et al.: Empirical validation of object-oriented metrics for predicting fault proneness models. Softw. Qual. J. **18**(1), 3–35 (2010)
28. Haykin, S.: Adaptive filters. Signal Process. Mag. **6** (1999)
29. Teng, S., Du, H., Wu, N., Zhang, W., Su, J.: A cooperative network intrusion detection based on fuzzy SVMs. J. Netw. **5**, 475–483 (2010)

Pandemic Effect on Education System Among University Students

Farzana Yesmin[1], Md. Mosfikur Rahman[1(✉)], Mohd. Saifuzzaman[2], and Nazmun Nessa Moon[1]

[1] Daffodil International University, Dhaka, Bangladesh
yesminfarzana.cse@gmail.com, mdmosfikurrahman.cse@gmail.com, moon@daffodilvarsity.edu.bd

[2] BJIT (Bangladesh Japan Information Technology) Limited, Dhaka, Bangladesh
mohdsaifuzzaman.cse@gmail.com

Abstract. Online exam systems are pieces of software used in online education to evaluate students' performance. However, both the security of the system they respond to and a broad understanding of the risks they may face must be addressed. In this article, we will look into the most recent issue, which is online exams and their security systems. Because most students depended on Google, their friends, and other resources for help, we focused on software-based security and online support for cheating like IP address tracing, block local file accessing etc. We did this project step by step like collecting data, data pre-processing, data cleaning, training data and trying to make a prediction of how the student cheated the most. Then we can start our future work as we want to increase the security of online exams. According to the findings of a few study papers we read, individuals that use e testing do so for all main kinds of assessment, including quizzes, midterms, and final exams. The most common format for these tests is multiple-choice, which is certainly the easiest to cheat on, but the faculty admits they aren't routinely employing anti-cheating tactics. So, this research paper will help the new world in the field of online exam systems.

Keywords: E-learning · Online exam · Machine learning · KNN · Naïve Bayes · Logistic regression

1 Introduction

Education has not been spared as technology has rapidly digitized all services and offerings, bringing them to the web platform. The pervasiveness of technology has facilitated a seamless shift to the online environment, thanks to ubiquitous high-speed internet and computers. Learning Management Systems, which are software applications that track, report, administer, and document materials shared with students, have advanced and adapted in universities, schools, and other educational institutions. This teaching and evaluation technique allows for a balanced approach to knowledge transmission as well as easy candidate grading. The goal of online exams is for assessment givers to be able to make the transition from offline to online processes. The online market is constantly expanding, and factors including convenience, scalability, expanded reach, and

L. Troiano et al. (Eds.): ICDLAIR 2021, LNNS 441, pp. 157–168, 2022.
https://doi.org/10.1007/978-3-030-98531-8_16

customization are influencing its adoption. Established evaluation methodologies are rapidly approaching saturation and will become obsolete in the near future. Exams are designed to fairly assess a student's ability. The exam results would be erroneous in representing their true competence if the exam's impartiality was compromised. Cheating raises various difficulties that we as an educational institution must be aware of, in addition to the incorrect assessment of students' learning progress. It's no longer necessary to scribble math formulas on your palms. Nowadays, students have devised ingenious techniques to cheat, most of which take advantage of the technology available to them. Sharing student's screen with another computer, utilizing cutting edge technological devices, using mobile apps and taking notes on a smartphone, creating false identities in order to obtain third-party assistance etc. Exam proctoring is the most prevalent solution. It is specifically developed to allow students to take online tests safely, with features like automated monitoring, ID verification, and machine lockdown, as well as IP address tracing and blocking local file access. This option, however, may be fraught with technical difficulties and significant expenditures. But, in order to discover a solution, we must first determine which elements are appropriate for pupils to cheat on. In this study work, my team and I used machine learning to forecast how students cheated the most and how they could improve their grades. We use three different sorts of algorithms in our project. To accurately train our data, we used Logistic Regression, KNN, and Naive Bayes. We test 20% of the data first, then train the remaining data our project's core theme is fairly straightforward. The title encapsulates the entire viewpoint. Though it is a recent issue, we need to enhance the security mechanism of online tests. Students employ a variety of strategies to pass. Students, on the other hand, constitute a country's backbone. As a result, we must give it considerable consideration. As a result, our initiative first identifies a problem before attempting to address it. As previously stated, it is a new trend, so just a few people are beginning to work with it. As a result, we were unable to locate any datasets suitable for our investigation. We manually collect data using a Google form. We put certain qualities there, and individuals entered data using these attribute questions. We gather approximately 1,000 data points, transform them to numerical values, train them, and then write our research using the trained data. Better and safer online exams would be provided by a secure online examination system. The whole process and quality of the exam process can be hampered by online examination systems that lack sufficient information security protocols. Our future work has a high level of ambition. We create a low-cost, practical solution that will benefit all types of students and educational institutions.

2 Review Works

Giustolisi, R. et al. (2013) aims to provide a research roadmap in the field. Safe, verifiable, and accountable e-exam systems have identified three separate study goals, each of which is led by particular research issues. They've just suggested a safe online examination method there [1]. Shraim, K. et al. (2019) despites the fact that higher education institutions in Palestine are increasingly using online exams, the perspectives of students on their use remain unexplored. As a result, the major focus of this study is to look into the online testing processes used by higher education institutions from the perspective of students who have taken online 2 exams at Palestine Technical University-Kadoorie (PTUK). Its findings will assist the institution in identifying essential and strategic components of effective E-exam design and administration to support students in higher education learning [2]. Muzaffar, A. W. et al. (2021) develops a review procedure that entails six phases according to the standard SLR standards. The first two phases are already addressed in the introduction to the article, and the next four phases are detailed in the accompanying segments [3]. Sarrayrih, M. A. et al. (2016) creates a completely automated Online Exam System that can gather, combine, and analyze data while also evaluating the program's impact. The system will only be accessible to authorised users with certain duties (Administrator: maintains the website, Institutes: register to administer examinations, Students: give exams online) [4]. Ramu, T. et al. (2013) states because of the high stakes involved in online exams, reliable student authentication is highly important. Authentication verifies that the users are who they claim to be. Authentication is used to verify the identity of online students and is a critical component of security in an online test setting [5]. Alotaibi, S. et al. (2010) to protect templates against fraudulent use, he recommended utilizing a distorted signal or feature vector. Using data hiding and watermarking techniques, fingerprint images can be made more secure by detecting changes and hiding one biometric within another. To protect against intruders, the scanned fingerprints will be stored in encrypted form. However, the time between fingerprint scans may be so short that it creates storage issues for such a large amount of data. As a result, the time interval can be increased to at least three seconds, allowing for a reasonable reduction in data volume [6]. Feinman, Y. et al. (2018) according to this research article, a quasi-experimental study was conducted with the primary objective of determining the effectiveness of nonbiometric security measures. Using a taxonomy of cheating reduction approaches based on the fraud triangle hypothesis, the security measures were selected for use. According to this study, the fact that the number of students who replied to all test items was higher on the proctored tests than on the unproctored exams may imply that students tend to skip more questions in unproctored contexts than they do in proctored environments Despite this, the sample was varied in terms of majors and included typical community college students. In addition, the sample of students in the study was not drawn at random from the population of all community college students who take introductory statistics in web-assisted environments. Whatever the case, the application of the security methods employed in the study may make it possible to assess student knowledge in a credible, low-cost, and convenient manner [7]. Apampa, K.M. et al. (2010) investigation into the current user security paradigm and the vulnerabilities that it contains is the subject of this study. According to the authors, a paradigm for user security in summative electronic evaluations has been developed specifically for this situation. Data security, location security, software security, and the network security of summative e-assessments are some of the issues to be concerned about. Instead of categorizing all impersonation threats as Type A, Type B, or Type C, the authors of this research categorize them into three types: A, B, and C. The tracking module computes the difference between two video frames in order to determine the location and position of the learner on the track. Student risk levels are determined at various points during a test using information from the tracker, which is fed into a classifier module. The current model's identification and authentication security goals are insufficient in terms of protecting student information from impersonation attacks [8]. Lin, N.H. et al. (2004) above study analyses how tool creation can work in conjunction with security to provide a learning environment that is safe, efficient, and effective. They detailed various different e-learning apps and categorized them according to their functions, which they then listed. They stated the security and privacy needs in terms of the situations and the types of data shared, based on this categorization. In addition, they planned to upgrade our e-learning systems in order to make them comply with the new distance education standard, known as the Sharable Content Object Reference Model (SCORM). They worked on student performance evaluation, Course Authoring and Student Assessment Application, Distributed Multimedia Presentation Application with Floor Control, Security and Privacy: Requirements and Solutions, Copyright Protection, and Floor Control Security while at this location [9]. Weiner, J. A. et al. (2017) aim of the research seems to be to compare online remote proctoring of high-stakes licensure examinations to traditional test center proctoring of the same examinations in order to determine whether or not these alternative test administration conditions are equivalent or whether or not they have potential differences. The findings of the study suggested that examinee scores achieved under online kiosk-based proctoring conditions were psychometrically sound and equivalent to the identical exams taken under traditional test center proctoring circumstances. Furthermore, examinees assessed online kiosk-proctored examinations well, and ratings of testing conditions were essentially uncorrelated with exam performance, implying that examinees' experiences under the various testing conditions had no impact on their performance on the exams. Globally speaking, the study's findings endorsed the use of a kiosk-based remote proctoring approach for high-stakes test delivery that was equivalent to using an onsite test center proctoring method. Despite this, there was no indication that remote proctoring had any effect on test results [10]. Al-Saleem, S. M. et al. (2014), in this study, author present a variety of security concerns in computer-based and online testing. In addition, they investigated a variety of authentication techniques for examinee authentication. Furthermore, they suggested a novel authentication system that combines the username/password with palm-based biometrics. The proposed scheme will enhance the authentication process up to an optimum level during computer-based or online-based testing. The video capturing technique should also be merged with the existing system, in order to make it more secure in the future. Also, the CATSIM simulation tool should be used to simulate the entire authentication and verification process. The suggested system should improve the authentication procedure during computer-based or online testing to an optimal level. The video capture technology should also be integrated into the existing system in the future to make it more secure. In addition, the CATSIM simulation tool should be utilized to model the whole authentication and verification process [11]. Jung, I. Y. et al. (2009) this author shows how the SeCOnE system provides both secure online exam management and an e-monitoring method for the prevention and detection of cheating. The strategies described in this article for preventing and detecting cheating encompass cheating methods known for the online test procedure through computer or Internet, although they may not address all conceivable cheating methods. This paper is intended for tests conducted over the Internet at a defined time with a single problem set, but with no restrictions on testing location. The fact that the second can be applied to exams given at various periods is a strong feature. To prevent cheating during the test, the examiner should prepare as many problems sets as there are exam hours. One overhead expense for this system is the preparation of the equipment should be used to monitor and authenticate the entities, such as Webcams and microphones. Another overhead to consider is the net burden caused by monitoring data transport and storage, but this is not a big issue when data compression is utilized and more monitor servers are provided. They should look at the consequences of hostile entities in the system, as well as the procedures involved in dealing with failures to improve this paper. The author should do some more work for improving this paper for enhancing security for online exams using group cryptography [12]. Mathapati, M. et al. (2017) this proposed technique increases the level of security for conducting online tests. The primary concept of this approach is the incorporation of live video and real tokens held by the user. This study demonstrates the viability of the results by testing a fully functional model. Through an evaluation of time, porn rate, and effort, it is demonstrated that the novel technique performance corresponds to the achievement level in traditional. The new approach is strong since it can survive any unfavorable circumstance of severe testing. The new method's performance is compatible with the standards and delivers steady, consistent results, making the new graphical password scheme reliable. When compared to previous approaches, the data show that the new methodology considerably enhances confrontation to attacks. Thus, they should improve their paper as well as system to meet the criteria of online examination systems. Finally, the new system secures and ensures the highest level of security during examinations. It can be improved by using compound character and image size [13]. Urosevic, A. et al. (2019) the aims and research question were justified and discussed throughout the literature review, and the findings were afterwards compared and contrasted to the findings from the semi-structured interviews. The research topic is: How can institutions assure proper student identity when examinations are taken outside of school by using various authentication methods? was examined using a variety of authentication techniques Institutions must keep up with what is going on across the world, which is justified by the rapid advancement of technology and potential applications of machine learning and artificial intelligence in e-learning. As observed throughout the literature study, potential outcomes and online tests have previously been introduced; however, security safeguards and fully supervised services may be lacking. Students need test time to focus on what they're doing, not to switch off and deal with any issues with third-party monitoring providers, for example. Unethical behavior is a problem that has been handled at institutions where tests are held on-site and hence does not pose a greater risk while working online. They conducted an empirical study on a group of online students from local and overseas Universities. The result shows the impact of questions type on the usability, in particular the amount of time taken by the introduction of the proposed approach. They should conduct a post experiment survey to collect student feedback on the proposed technique [14]. Lin, J. C. W. et al. (2016) A plethora of methods has been developed in this paper to anonymize relational data. However, preventing the leakage of sensitive information in transactional data is also a key concern. Algorithms suggested for anonymizing transactional data often result in significant information loss. In this work, they introduce PTA, a unique anonymization system composed of three modules. It anonymized transactional data with little information loss and is also highly fast when compared to state-of-the-art transactional data anonymization techniques. Significant tests should be conducted to evaluate the performance of the developed system against state-of-the-art techniques for anonymizing transactional data in terms of runtime and information loss. The outcome of this work is quite impressive and also it outperforms the existing handcrafted feature extraction methods. It is obvious that integration of different approaches will improve the performance of the whole system within the open-source software development Moodle [15]. Shetu, S. F. et al. (2021) shows an e-learning model to increase academic performance of students using a platform. In their paper, they tried to highlight the points regarding the pandemic affected student's academic performance [16] (Fig. 1).

3 Research Methodology

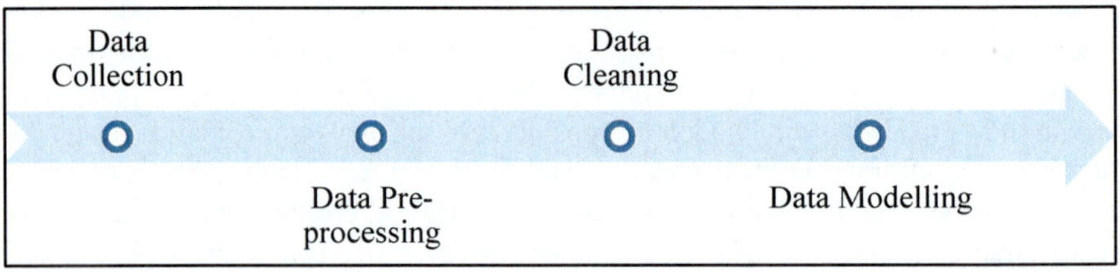

Fig. 1. Research methodology

3.1 Data Collection

The authors are making use of the real-time data set that was acquired through a Google form on their website. The form was distributed on social media, and from there, all of the necessary information was gathered. The dataset contains information about students' online exam experiences, including whether or not they received any online assistance. There were some questions about the dataset. Questions are:

- You're Grade in Online Examination System.
- Get help in exams from your classmates.
- Get help in exam from Google and YouTube
- Friend circle grouping exam
- Get help from any other source?

Answer to the first question was two; the first one was increased, and the second one was de-creased, resulting in a total of two answers. The responses to the second question were given in percentages. The answer to the third question is the same as the answer to the second question. The fourth and fifth questions have two possible answers: yes and no. The final question had four responses, which were: electricity problem, network problems, Device Management problem, and others.

3.2 Data Pre-processing

Following the collection of the data, the dataset was discovered to be in Microsoft Excel format. For the purposes of this project, data was first converted into a CSV format. Many nominal values were discovered after the conversion process was complete. Those nominal values are manually processed and converted to numerical values by a computer program. These values were then translated into integer numbers using a string conversion algorithm. Take, for example, the following:

- In Result
  - Increase: 1
  - Decrease: 0
- Online Help and Friend Help
  - Percentage deviation from typical numerical values.
- Group Exam and other source
  - Yes: 1,
  - No: 0
- Problem During exam
  - Electricity Problem: 1
  - Net Problem: 2
  - Device Management Problem: 3
  - Others: 4

3.3 Data Cleaning

Following the cleaning of the data, we proceed to the identification of the dependent and independent variables. In this particular instance, we have chosen X and Y. Yes is the dependent variable, whereas X is the independent variable in this situation. Following that, we selected 20% of the data to train on and 80% of the data to give accuracy for the model.

3.4 Data Modelling

3.4.1 K-Nearest Neighbor

KNN is an abbreviation for K-Nearest Neighbors, which is a type of proximity search. A learning algorithm is used, and it is under the supervision of a trained individual. It establishes the distance between each row of training data and each row of test data using the distance function. By their respective values, arrange the distances determined in ascending order. The first k rows of a sorted array are returned by this function. In this row, find out which class is the most frequently seen. Should be returned as the projected class.

$$D(x_i, x_j) = \sqrt{(x_i - x_j)^2 + (y_i - y_j)^2} \ldots\ldots\ldots \quad (1)$$

3.4.2 Logistic Regression

Machine learning algorithm logistic regression can be used to predict if an observation belongs to one of two classes based on its characteristics. In financial analysis, linear regression is a machine learning technique based on supervised learning. It's in charge of carrying out a regression operation. These models use independent variables to predict the value of a prediction.

The sigmoid function has the following formula:

$$\text{Sigmoid}(x) = \frac{1}{1 + e^{-x}} \ldots\ldots\ldots \quad (2)$$

Therefore, for Logistic Regression the equation becomes equation (iii).

$$P = \frac{1}{1 + e^{-(b' + b'' x)}} \ldots\ldots\ldots \quad (3)$$

3.4.3 Gaussian Naïve Bayes

A simple learning algorithm, Naive Bayes makes use of the Bayes rule in conjunction with a strong assumption that the attributes are conditionally independent, given the class in which they are used. It is possible to build a Bayesian classifier using a probabilistic model in which the classification is represented by a latent variable that is probabilistically related to the observed variables.

$$P(x_i | y) = \frac{1}{\sqrt{2\pi\sigma_y^2}} \exp\left(-\frac{(x_i - \mu_y)^2}{2\sigma_y^2}\right) \ldots\ldots\ldots \quad (4)$$

4 Result and Discussion

4.1 Confusion Matrix

The values of the KNN, Logistic, and Naive Bayes Confusion Matrix are listed in this table. Here the table of this Confusion Matrix is True Positive 3 (TP), True Negative (TN), False Positive (FP) and False Negative (FN) is given in Table 1.

Table 1. Confusion matrix

Algorithm		Increased	Decreased
KNN	Increased	174	14
	Decreased	6	4
Logistic regression	Increased	178	0
	Decreased	22	0
Naïve Bayes	Increased	178	0
	Decreased	22	0

4.2 Classification Report

A Classification report must be created to evaluate the accuracy of the predictions provided by a classification algorithm. Classification metrics precision, recall, and f1-score are displayed in a single report row for each class. Metrics like true positives and false negatives are factored into the formula for calculating performance. A comparison table has been created with both the decreased and raised values. The algorithm's precision is also taken into account. Table 2 shows the classification report.

Table 2. Classification report

Algorithm	Class	Precision	Recall	F1-Score	Accuracy (%)
KNN	Decreased	0.80	0.73	0.76	95
	Increased	0.97	0.98	0.97	
Logistic regression	Decreased	0	0	0	89
	Increased	0.89	1	0.94	
Naïve Bayes	Decreased	0	0	0	89
	Increased	0.89	1	0.94	

4.3 Accuracy

The machine learning model measures the accuracy with which a model finds correlations and patterns between variables in a dataset based on the input data, or training data. This accuracy chart illustrates the relative importance of KNN, Logistic, and Naive Bayes methods. And KNN had the highest Accuracy rating, earning a score of 95%. Figure 2 shows the accuracy chart of the used machine learning algorithms.

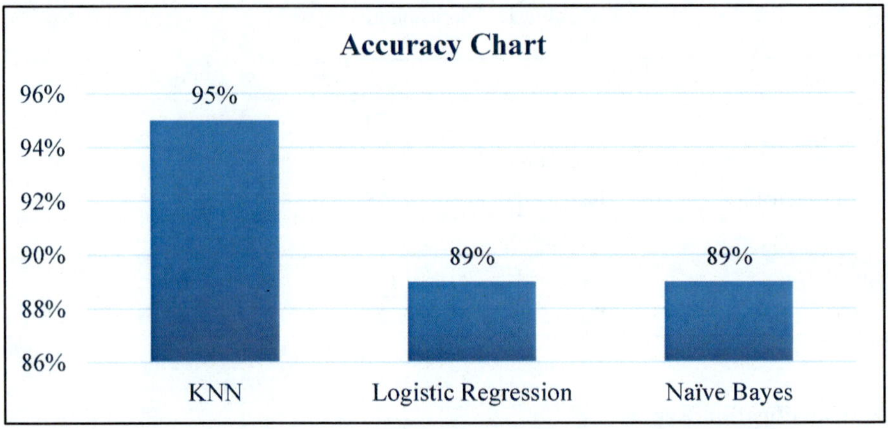

Fig. 2. Accuracy chart

5 Comparative Analysis

An online examination system makes it possible to prepare and deliver exams entirely online. Because of internet-based configuration options, this is a real possibility. This paper's goal is to make the online test system more secure. K Nearest Neighbors, Logistic Regression, and Gaussian Nave Bayes were a few of the confusion matrices we used in our research. The classification report is summarized in the table below (ii). On a 95% of all occasions, the KNN matrix performed flawlessly. There were two methods used to achieve the same level of accuracy: logistic regression and Gaussian Naive Bayes. The precision of the results used by different authors and researchers, on the other hand, has led to conflicting conclusions. Online exams can be made more secure by utilizing biometric authentication methods such as fingerprint recognition. Through the use of online testing methods, the objectivity of evaluations can be improved while the effort required to make corrections is reduced. For new student orientation and registration, this is very useful when utilized in large classrooms (STEOP). Modern assessment appears to be essential for modern education at this time. The data for this report was gathered using a google form. Often times, authors may gather their information from a variety of various websites all over the internet. This leads me to believe that updating exam security using real data is preferable in this instance. Students who use e-testing for quizzes, midterms, and exams of all kinds, according to research after research. However, instructors report they do not regularly use procedures to avoid academic dishonesty despite the most common exam format being multiple choice, which makes cheating the easiest. The WebCT courseware platform offers numerous preventative measures for educators that are as easy to implement as using good judgment and selecting a single directive. If academic dishonesty is an issue, more should be done to prevent it for obvious reasons. There isn't a solution in adding more work to the faculty; rather, swift decisions about basic options and responsibilities must be made.

6 Conclusion and Future Work

This paper explains how our online test system will be safer and more secure, as well, as how students might improve their grades by cheating online. The measures presented in this work for preventing and detecting cheating cover cheating methods identified for online exam processes through computer or Internet, however, they may not cover all possible 95% 89% 89% Accuracy Chart KNN Logistic Regression Naïve Bayes 4 cheating methods. Our primary goal is to improve the examination atmosphere. We will take things one-step at a time. To begin, this study will investigate how students cheated, as well as the difficulties they encountered during the online exam. Following that, we will address their concerns as well as the subject of cheating. As a result, the online testing method improved day by day. We concentrated on software based security and their internet assistance for cheating because most students relied on Google, their friends, and other resources for aid. That is why we concentrate mostly on that aspect. To classify our paper's intention, we employ KNN, Logistic Regression, and Naive Bayes algorithms in our coding section. In computer-based and online-based testing, we have looked into many security issues. We have also looked into a number of other authentication mechanisms for examinees. Finally, the lack of in-person supervision owing to geographical limits, along with accessible technologies, has made it easier for students to cheat on online tests. Knowing how students cheat in novel ways can help you find the proper answer and alleviate security concerns. In the future, we intend to improve our projection by incorporating more and deeper records from our own database, which we are continuously building with increasingly complex data. We would like to incorporate more security solutions, as well as how to create them from a Bangladeshi standpoint. We currently do not have enough datasets; but, in the future, we will have a large dataset with many variables that will indicate how we should improve security in the online examination system. The consequences of hostile entities in the system, as well as the mechanisms involved in handling errors, will be the focus of future research.

Acknowledgment. First and foremost, we would like to express our profound gratitude to Allah for His tremendous grace, which has enabled us to complete this work successfully. We also express our gratitude to my respectable course instructor. Because he is with us throughout our whole journey. Finally, we'd want to thank everyone who helped us, including well-wishers, friends, family, and elders. In addition to concepts. This study is the result of a great deal of hard work.

References

1. Giustolisi, R., Lenzini, G., Bella, G.: What security for electronic exams? In: 2013 International Conference on Risks and Security of Internet and Systems (CRiSIS), pp. 1–5. IEEE (2013)
2. Shraim, K.: Online examination practices in higher education institutions: learners' perspectives. Turk. Online J. Distance Educ. **20**(4), 185–196 (2019)
3. Muzaffar, A.W., Tahir, M., Anwar, M.W., Chaudry, Q., Mir, S.R., Rasheed, Y.: A systematic review of online exams solutions in e-learning: techniques, tools, and global adoption. IEEE Access **9**, 32689–32712 (2021)

4. Sarrayrih, M.A.: Implementation and security development online exam, performances and problems for online university exam. Int. J. Comput. Sci. Inf. Secur. **14**(1), 24 (2016)
5. Ramu, T., Arivoli, T.: A framework of secure biometric based online exam authentication: an alternative to traditional exam. Int J Sci Eng Res **4**(11), 52–60 (2013)
6. Alotaibi, S.: Using biometrics authentication via fingerprint recognition in e-exams in e-learning environment (2010)
7. Feinman, Y.: Security mechanisms on web-based exams in introductory statistics community college courses. J. Soc. Behav. Health Sci. **12**(1), 11 (2018)
8. Apampa, K.M., Wills, G., Argles, D.: User security issues in summative e-assessment security. Int. J. Digit. Soc. (IJDS) **1**(2), 1–13 (2010)
9. Lin, N.H., Korba, L., Yee, G., Shih, T.K., Lin, H.W.: Security and privacy technologies for distance education applications. In: 18th International Conference on Advanced Information Networking and Applications, 2004. AINA 2004. Vol. 1, pp. 580–585. IEEE (2004)
10. Weiner, J.A., Hurtz, G.M.: A comparative study of online remote proctored versus onsite proctored high-stakes exams. J. Appl. Test. Technol. **18**(1), 13–20 (2017)
11. Al-Saleem, S.M., Ullah, H.: Security considerations and recommendations in computer-based testing. Sci. World J. **2014**, 1–7 (2014)
12. Jung, I.Y., Yeom, H.Y.: Enhanced security for online exams using group cryptography. IEEE Trans. Educ. **52**(3), 340–349 (2009)
13. Mathapati, M., Kumaran, T.S., Kumar, A.K., Kumar, S.V.: Secure online examination by using graphical own image password scheme. In: 2017 IEEE International Conference on Smart Technologies and Management for Computing, Communication, Controls, Energy and Materials (ICSTM) pp. 160–164. IEEE (2017)
14. Urosevic, A.: Student authentication framework for Online exams outside of school (2019)
15. Lin, J.C.W., Liu, Q., Fournier-Viger, P., Hong, T.P.: PTA: an efficient system for transaction database anonymization. IEEE Access **4**, 6467–6479 (2016)
16. Shetu, S.F., Rahman, M.M., Ahmed, A., Mahin, M.F., Akib, M.A.U., Saifuzzaman, M.: Impactful e-learning framework: a new hybrid form of education. Curr. Res. Behav. Sci. **2**, 100038 (2021). https://doi.org/10.1016/j.crbeha.2021.100037

Prediction of Migration Outcome Using Machine Learning

S. M. Rabiul Islam(✉), Nazmun Nessa Moon(✉), Mohammad Monirul Islam, Refath Ara Hossain, Shayla Sharmin, and Asif Mostafiz

Daffodil International University, Dhaka, Bangladesh
{rabiul25-917,refath.cse,shayla.cse}@diu.edu.bd, {moon, monirul}@daffodilvarsity.edu.bd

Abstract. The decision to migrate is mainly influenced by a combination of factors such as income, security, politics, lifestyle, education, and environmental pressures. Problems faced by Bangladeshi migrants include finding an appropriate matching job, lack of information on migration opportunities and risks; discrimination, exploitation, and abuse while overseas; and insufficient services. Traditional migration model as such gravity and radiation model based on population and distance features focused. These models are not enough for complicated migration dynamics. It is challenging to identify appropriate migration destinations for individuals since demand and supply change rapidly. We prepared a machine learning model with survey data where 1000+ individual migration worker participants from 10+ countries (the major countries where Bangladeshi workers migrate). Then applied decision tree and random forest classification to predict the successful migration outcome. The result shows that the decision tree model produces more accuracy than a random forest with this small data set. However future work should continue with large data set to explore potential machine learning techniques to find appropriate migration destination where to remain migrant satisfied and happy with the job.

Keywords: Human-migration · Decision tree · Machine learning · Random forest classifier

1 Introduction

Migration prediction is a type of system where it is possible to know what will happen after migration upfront through the knowledge of existing migration. The decision for migration is vital and critical for individuals, family as well country however due to lack of information, knowledge, and tempering information from agencies, many people in Bangladesh get trapped and bear lifelong suffering. By asking some simple questions about candidates like Gender, age, Trade/Educational qualification, Skill, language-skill, Cost of migration, Destination-Country, etc. this system can predict if the candidate will be happy for the opportunity after migration.

The reason behind migration and complexity and varies based on different parameters, the weather introduces a lot of different types of threats to humans [1]. For low-lying

L. Troiano et al. (Eds.): ICDLAIR 2021, LNNS 441, pp. 169–182, 2022.
https://doi.org/10.1007/978-3-030-98531-8_17

coastal communities faces difficulty in life in a variety of form which includes severe storms, torrent, precipitation, and rise of sea-level [2]. The livelihood opportunities are affected by the combination of environmental pressures and climatic change together [3]. Changes of place are common for untoward environmental changes [4]. The risk and perception on migration research [5, 6], some research found more specific patterns with migration and climate [7]. Factors such as gender balance, equality [8, 9], privacy, safety, and legal concern [10], and lifestyle, quality of life [11], also influence migration.

The fact of the reason behind the migration is complex and changes continuously. So, the gravity and radiation model can help to predict migration patterns that are not enough for individual migrants to avail the opportunity rather need a proper systematic guideline where is the best matching place to migrate considering age, sex, skill, education, language skill, etc. With the help of the machine learning technique, it is possible to prepare a model to predict migration satisfaction precisely. By asking few simple questions to potential migration workers, the machine will respond how the outcome will be, will it be a success or failure.

2 Review Works

The Big data perspective of human migration and return of migrants has been covered by Sîrbu1, Andrienko, Andrienko, Boldrini, Conti, Giannotti, Guidotti, Simone Bertoli, Kim, Muntean, Pappalardo, Passarella, Pedreschi1, Pollacci, Pratesi and Sharma (2020). They provide a detail analysis of different phases of migration, comparing with traditional migration model and show example how big data can play role in integration of migrant in destination country as well as the pattern of return migrations. Data driven analysis at higher spatial resolution on retail data such as by analyzing mobile call record, food habit, travel history and social media data, it's possible to develop a sentiment model of immigrant [12].

Christian M. Alis, Erika Fille Legara, and Christopher Monterola (2021) show the improvement of radiation model where migration decision based on population, attractiveness of distance and neighboring locations. However, in their study more amenities are added like offices, schools, holiday places beside traditional model and reported a relative improvement of 7.7% in mean absolute percentage error [13] by using census data of Philippines. Hussain, Anuar, Mohd, Masrom and Ahmad (2021) show the reverse migration pattern can help for better planning of rural and urbanization as well as handle pandemic situation. In the reverse migration model literature review, they reviewed three machine learning technique decision tree, random forest and linear regression. Compared to traditional gravity and radiation model, this machine learning technique provides more accurate result which can be helpful for human forecasting and planning [14]. The paper [15], detect pattern from raw data after applying machine learning technique and present the pattern in visualization tool. Knowledge Discovery Pattern Recognition and Data Visualization combined for by analyzing and to discover collective migration patterns. Showed the stages of KDD build the database of indicators country wise. From visualization perspective while change different parameter from that and adjust the output which giving extended view of primary analysis such as countries with greater GDP and significant percentage of internet users; most of the migration took place among those destination. Best, Gilligan, Baroud, Carrico, Donato, Ackerly and Mallick (2020) performed random forest analysis on two survey to predict the migration pattern of Bangladesh. It shows for complex migration data model random forest outperform logistic regression and support vector machine to predict migration cases. Show the different reason of migration like education, environmental, health, trade, job, etc. Though random forest classifier can useful for complex data model still need support from traditional regression methods to identifying attributes correlation with migration [16]. Robinson, Dilkina, Moreno-Cruz (2020) in paper title "Modeling migration patterns in the USA under sea level rise" is an experiment that predict the future migration pattern of USA due to change in sea level. This study considered the previous similar changes due to change of sea level over different country and how it effects the livelihood due to the change. This experiment shows the integration of climate change and human migration model and derive an extended model that can be useful for forecasting [17].

Though all these papers mostly recognized the pattern why people migrate or while return to origin country however the study missing of migration satisfaction. People want to move for lot of different reason, but they want to bring positive change in life. In many cases this notion of creating a positive future fails due to lack of information about prospect and matching of individual attributes.

3 Research Methodology

This part explains the methodology and procedure for the research. In addition, tools, the way of collecting data, data processing, analysis of outcome from statistical perspective and implementation.

3.1 Research Subject and Instrumentation

A research subject is about research area and understanding its different parameters. We notice that information is the fundamental piece of the examination. It is an exceptionally basic part for a specialist to discover amazing data and wonderful algorithm or model for our exploration work. We likewise need to examine related examination papers.

3.2 Data Collection Procedure

Data used in this project are collected from various group of people who are currently working in other country. Initially we thought to collect data from different agency who work with migration but due to lockdown we could not execute initial plan. Through our friends, family and friends of friends in different country helped us to collect data. A online survey was used to capture data.

The research used 1000+ data which have been collected through online survey that designed with attributes which determined the feature class. Among those records the number of males is 874 (85.18%) and the number of females is 152 (14.82).

3.3 Data Pre-processing

Data set contains 1026 records in which some values are missing. In Fig. 1, we have handled the missing values by using Null and Mean imputation. It is a method in which the missing value on a certain attribute is replaced by the mean of the available cases. This method maintains the sample size and easy to use. Figure. 1 show the heatmap function output of transformed raw data.

```
sns.heatmap(dst.isnull())

<matplotlib.axes._subplots.AxesSubplot at 0x7f90f91a3a50>
```

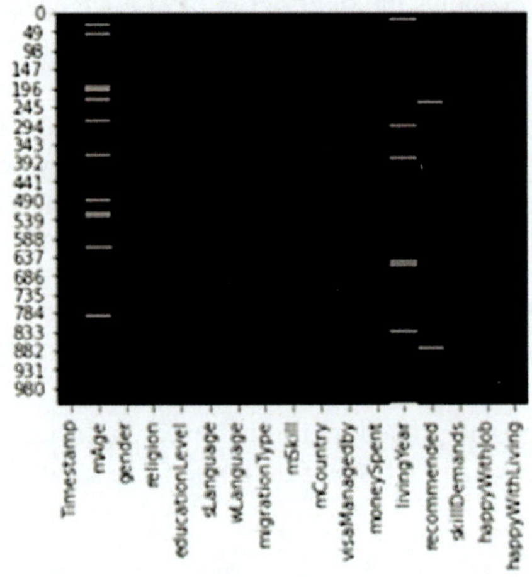

Fig. 1. Missing data represent

In data there were instances where some of feature column value missing. Following Fig. 2 show the code snippet used to handle missing data.

```
dst["livingYear"]=dst["livingYear"].fillna(dst["livingYear"].mean())
dst["recommended"]=dst["recommended"].fillna(dst["recommended"].mode()[0])
dst["mAge"]=dst["mAge"].fillna(dst["mAge"].mean())
```

Fig. 2. Code snippet to handle missing data

Figure 3 show the output of heatmap with null check, it appears no more null and data in a state can be used for preparing machine learning model.

```
sns.heatmap(dst.isnull())

<matplotlib.axes._subplots.AxesSubplot at 0x7f90fa8d1210>
```

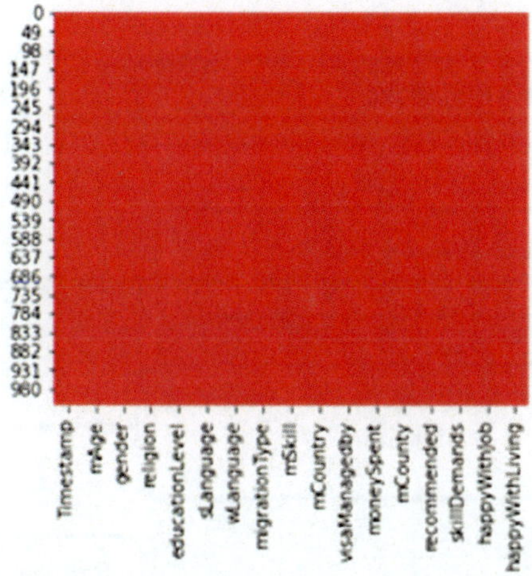

Fig. 3. After processing missing data.

3.4 Attributes and Feature Selection

In the survey used 19 different attributes along with output class. Attributes: the main attributes are (Age, Gender, Religion, Highest Level of Education, Language speaking skill, Language writing skill, Migration Type, skill at time of migration, Migrated Country, Visa Sponsored Type, Money Spent for migration, Number of years live in migrated country, Same Country recommended, Preferred skill for this country, Happy with Job/Business, Happy staying there). Moreover, data divided to two separates data sets "Training data set to build the model and test dataset to evaluate the model".

The dataset contains 19 attributes though all those are not used for decision tree node. Generated a correlated logic first then consider the attribute which are important only. However, for random forest model all attributes used. For model and training purpose data have been divided into two groups.

3.5 Algorithm for Predicting Migration Satisfaction

The data captured for attributes are non-linear and the class field is binary. So, need a binary classifier algorithm which can pop result for new record based on training. For binary classification two most popular algorithms are decision tree and random forest classifier. The Fig. 4 depict the steps from data collection to final result finding and analysis.

Using survey collected 1000+ records, a subset of records has been showing in Table 1 where only 6 feature columns shown those are used to develop the decision tree to decide the output column. Though 19 Attributes collected after checking the data pattern and optimization of features, we simplified the attributes for decision tree building. However, for random forest classifier all the 19 attributes have been used.

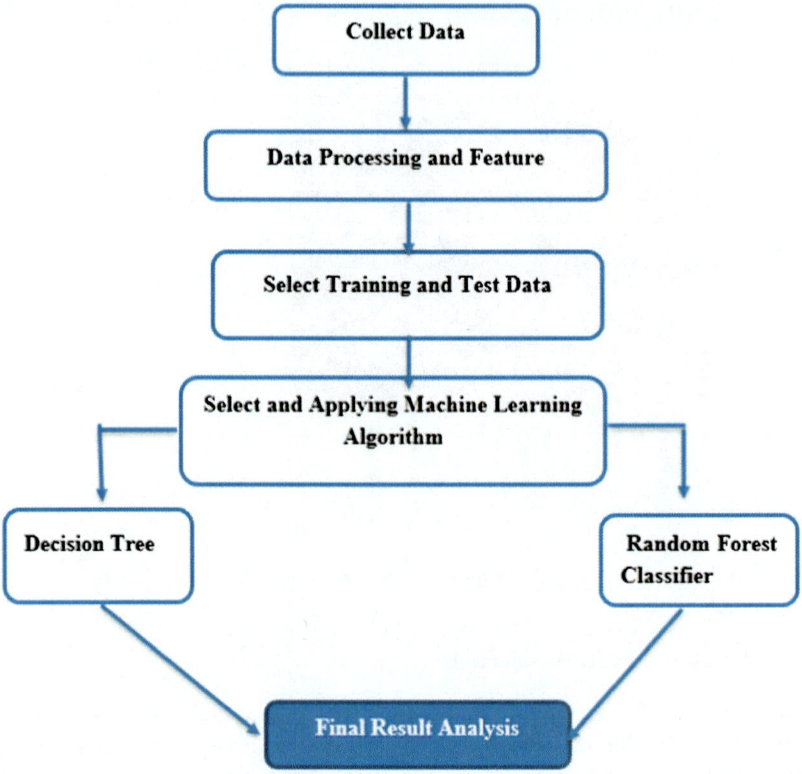

Fig. 4. Proposed model structure

Table 1. Sample normalized attributes show

Sl	Age	Gender	Skill Type	Visa Sponsored By	Skill	Target Language Skill	Happy with Job
1	51	Male	Unskilled	Agency	Cook	No	No
2	47	Male	Skilled	Self	Teacher	Yes	Yes
3	37	Female	Semi-Skilled	Agency	Nurse	Yes	Yes
4	42	Male	Unskilled	Agency	Construction worker	No	No
5	25	Male	Unskilled	Agency	Construction worker	No	Yes
6	30	Female	Skilled	Relative	Engineer	Yes	Yes
7	33	Male	Semi-Skilled	Agency	Driver	No	Yes
8	56	Male	Skilled	Self	Doctor	Yes	Yes
9	23	Female	Semi-Skilled	Agency	IT	No	No
10	27	Male	Unskilled	Relative	Driver	No	Yes